Computer-Based Electronic Measurement:

An Introductory Electronics Laboratory Workbook Based on *LabView* and *Virtual Bench*

A. Bruce Buckman
University of Texas at Austin

Prentice Hall
Upper Saddle River, NJ 07458

Library of Congress Cataloging-in-Publication Data

Buckman, A. Bruce
  Computer-based electronic measurement: introductory electronics laboratory workbook based on LabView and Virtual Bench / A. Bruce Buckman
  p. cm.
  ISBN 0-201-36182-5
  1. Electronic measurements--Computer stimulation. 2. LabView. I. Title.

TK7878.B83   1999
621.3815'48--dc21   99-049913

Publisher: Tom Robbins
Acquisitions editor: Alice Dworkin
Editor-in-chief: Marcia Horton
Director of production and manufacturing: David W. Riccardi
Executive Managing editor: Vince O'Brien
Managing editor: David George
Editorial supervision: Scott Disanno
Cover director: Jayne Conte
Marketing manager: Danny Hoyt
Manufacturing buyer: Pat Brown

 © 2000 by Prentice-Hall, Inc.
Upper Saddle River, New Jersey 07458

All rights reserved. No part of this book may be reproduced, in any form or by any means, without permission in writing from the publisher.

The author and publisher of this book have used their best efforts in preparing this book. These efforts include the development, research, and testing of the theories and programs to determine their effectiveness. The author and publisher shall not be liable in any event for incidental or consequential damages in connection with, or arising out of, the furnishing, performance, or use of these programs.
Printed in the United States of America

10 9 8 7 6 5 4 3 2 1

ISBN 0-201-36182-5

Prentice-Hall International (UK) Limited, London
Prentice-Hall of Australia Pty. Limited, Sydney
Prentice-Hall Canada, Inc., Toronto
Prentice-Hall Hispanoamericana, S.A., Mexico
Prentice-Hall of India Private Limited, New Delhi
Prentice-Hall of Japan, Inc., Tokyo
Prentice-Hall (Singapore) Pte., Ltd., Singapore
Editora Prentice-Hall do Brazil, Ltda., Rio de Janeiro

TABLE OF CONTENTS

Laboratory Measurements with *LabVIEW* and *Virtual Bench*

Chapter M1. Signal Generation	M1-1
Chapter M2. Waveform Capture	M2-1
Chapter M3. Measuring Waveform Spectral Content	M3-1
Chapter M4. Measuring Complex Transfer Function	M4-1
Chapter M5. Measuring Current and Impedance	M5-1
Chapter M6. Measuring Current-voltage (i-v) characteristics for Non-linear Devices	M6-1
Chapter M7. Small-signal Parameter Measurements	M7-1

Laboratory Projects

Amplifiers I: Op-amp realizations
Project 1. Non-inverting Amplifier	P1-1
Project 2. Inverting Amplifier	P2-1

Impedance Transformations
Project 3. Artificial Inductor	P3-1
Project 4. Transformers and Impedance	P4-1

Filters and Phase-shifters
Project 5. First- and Second-order Low Pass Filters	P5-1
Project 6. All-pass filters and Other Phase-Shifters	P6-1

Non-linear Circuits I: Diode Circuits
Project 7. Rectifiers	P7-1
Project 8. Limiting Circuits	P8-1

Oscillators
Project 9. RC Relaxation Oscillator	P9-1
Project 10. Phase-shift Oscillator	P10-1

Amplifiers II: Single-stage discrete devices
Project 11. Amplifier Circuits with Bipolar Junction Transistors (BJT's)	P11-1
Project 12. The Source-follower FET Amplifier	P12-1

Nonlinear Circuits II: Modulation
Project 13. Voltage-Controlled Gain and Amplitude Modulation	P13-1

Appendix. PSPICE Essentials	A-1

Author's Preface: Ways to Use This Workbook

1. Why Should Undergraduate Labs Go to DAQ?

This workbook grew out of the process of converting the introductory electrical and computer engineering laboratory course at the University of Texas at Austin to an approach based on digital signal acquisition (DAQ) instrumentation. There are a variety of reasons why an undergraduate lab might convert to DAQ, but the principal ones are these:

Emerging Dominance of DAQ Instrumentation

Industry and research laboratories along with undergraduate university laboratories have felt the pinch of rising costs of laboratory instrumentation. This trend created a strong market for laboratory instruments based on the personal computer. The appropriate software, operating through a plug-in digital acquisition (DAQ) board or card in the computer, can create the virtual instrument equivalent of $30-80K worth of conventional electronics laboratory instrumentation with an order-of-magnitude cost reduction. Thus, it is far from surprising that these virtual instruments are rapidly replacing their free-standing counterparts in the industrial facilities that hire many electrical and computer engineering graduates. Since this type of instrumentation will soon be what the typical graduate will encounter on the job, undergraduate laboratory courses need to familiarize students with it.

Age of Installed Base of Instrumentation in Undergraduate Labs

Whether organized as separate courses or appended to lecture courses, undergraduate laboratories have typically been poorly equipped, as compared to state-of-the-art laboratories inside or outside of academe. This situation has worsened as the pace of improvements in laboratory education has accelerated. The age of the installed base of instrumentation in the typical undergraduate electronics laboratory also imposes additional costs such as a) increased technician demand to keep older equipment operating, and b) the de-motivating effect on students of having to use such equipment.

Ease of Use as a Student Motivator

Since virtual instruments reside in the computer, much of the drudgery formerly associated with laboratory work and the production of a report on laboratory results can be reduced or eliminated. Both *Virtual Bench* and *LabVIEW* provide the capability to produce and print graphical versions of tabulated data, oscilloscope traces, frequency spectra, transistor curve traces, and the like. Furthermore, the look and feel of the instruments in *Virtual Bench* is so similar to that of free-standing instrumentation that the software becomes a teaching tool for training students to use the available hardware instruments.

2. Using This Workbook in a DAQ-Based Undergraduate Lab Course

Undergraduate electronics laboratory courses are typically organized in one of two ways. Some schools offer independent laboratory courses, which tend to focus on measurement techniques and laboratory skills. Others offer laboratory sections attached to the various theory courses, which focus on illustrating concepts presented in the lectures, with the measurement and test skills needed to perform the experiments introduced on a "just-in-time" basis.

This workbook is designed to be used with either type of course. It is divided into a set of Measurement chapters, designated with the prefix, "M," and a set of Projects, designated with the prefix, "P." The following approaches are suggested for selecting and sequencing workbook material:

Lab courses emphasizing measurement and test skills. The lab instructor can select from the measurement Chapters 1-7 the measurements and methods to cover. Within the measurement chapters are a few Lab Skill Exercises for the students to do in the laboratory. Generally, these Lab Skill Exercises by themselves do not offer enough example material to teach each measurement skill thoroughly. The lab instructor can select for further examples tasks from the project chapters. For example, the Projects on amplifiers (P1, P2) and filters (P5, P6) contain many tasks that require measurement of transfer functions (Ch. M4), or transient responses (Ch. M2).

Lab courses emphasizing electronics concepts from associated theory courses. The lab instructor can select material that matches the desired electronics concept from the Projects. The Projects match the basic analog electrical and electronic functions covered in most texts. Within each Project task the workbook prompts the reader to choose a measurement or test method and directs the reader to the corresponding sections of the Measurement chapters. The choice of measurement method, when available, can either be left to the student or assigned by the instructor.

3. Using the Workbook with Other Courseware

For a successful laboratory experience, the student may need material from three areas for which this workbook has no space:
- Electrical and electronics basics,
- PSPICE simulation of circuits, and
- *LabVIEW* programming fundamentals.

Many good texts exist in all these areas, and this workbook is written to complement rather than compete with them.

The Projects contain reference tables linking each project to sections in the following electrical and electronics texts:
1. James W. Nilsson and Susan A. Riedel, *Electric Circuits* (6th ed.), Prentice-Hall (2000).
2. Adel Sedra and Kenneth C. Smith, *Microelectronic Circuits* (3rd ed.), Oxford University Press (1991).
3. Allen R. Hambley, Electronics (2nd ed.), Prentice-Hall (2000).

Instructors using other basic texts should be able to identify points of reference in such texts easily.

This workbook contains an Appendix that summarizes the bare essentials of the Windows version of PSPICE that are needed to perform the simulation tasks in each of the Projects. An excellent additional reference is:

Roy W. Goody, *MicroSim PSPICE for Windows* (Vol 1) 2nd ed. Prentice-Hall (1998).

The choice of software to use with a DAQ laboratory can be made a number of different ways. National Instruments' *LabVIEW* offers almost unlimited flexibility for the user but can require more time to teach basic programming concepts that are not needed if the executable package, *Virtual Bench*, is selected. *LabVIEW* has an OnLine Help utility within the programming environment that provides nearly all the information needed about individual virtual instruments. Two additional texts on *LabVIEW* make excellent references:

1. Robert H. Bishop, *Learning with LabVIEW*, Addison-Wesley (1998).
2. Lisa K. Wells and Jeffrey Travis, *LabVIEW for Everyone – Graphical Programming Made Even Easier*, Prentice-Hall (1997).

At the end of each measurement chapter in this workbook is a *LabVIEW* Reference Table that directs the reader to appropriate supporting reference material.

For the instructor trying to decide how much *LabVIEW* programming can be added to an already crowded curriculum, one final comment is in order. The workbook material that uses *LabVIEW* can be introduced to the students in the following ways:
1. They can be asked to build the measurement VI's from scratch, following the guidelines in the measurement chapters. This approach fosters the most *LabVIEW* expertise, but requires the study of some supplementary material on *LabVIEW* programming from the references.
2. The VI's in the measurement chapters can be provided to the students in finished form. All of the measurement VI's in this workbook are available for downloading from these internet addresses:
    - http://www.ece.utexas.edu/~buckman/
    - http://www.natinst.com/

No warranty as to fitness for a particular use is provided with the software programs, either as taken directly from this workbook or as provided by downloading from the internet. Different combinations of DAQ board and host computer will result in varying performance levels. Some of these combinations may even require minor modifications in the VI's in order to run them.

Although this workbook relies on three large, complicated, and versatile software products, *LabVIEW*, *Virtual Bench*, and PSPICE, it makes no attempt to explain all of the many features of these programs. It approaches learning a new piece of software the same way a new purchaser of a word-processor or spreadsheet would: Start with a small but interesting task, learn to do that, and then branch out. I learned *LabVIEW* this way in my own research.

Acknowledgments

In closing, I gratefully acknowledge the encouragement and assistance of Ravi Marawar, Lisa Wells, Debby Clarke, Ray Almgren, Liz Stice, and Steve Symons of National Instruments. I also thank the Department of Electrical and Computer Engineering at The University of Texas at Austin for rising to the challenge and seeing the opportunity of converting our introductory laboratory course to DAQ-based instrumentation. I also thank the many teaching assistants at The University of Texas who helped work the bugs out of the course as it was being introduced. Finally, I thank my wife, Carole, for doing whatever was required, from cheerleading to proofreading.

Austin, Texas
August 1999

Foreword: Electronics Laboratory Safety

There aren't many hazards in a well-equipped electronics laboratory, but you need to be aware of a few. The most important are personal hazards to you. Also important are hazards to delicate, expensive equipment.

Personal Hazards

First and foremost is the hazard of electrical shock. On a typical electronics board, voltage levels are unlikely to exceed 20 volts. **Don't let that low number give you a false sense of security.** Voltage level is not the key determiner of the severity of electric shock. The key elements in determining shock severity are current and time. Generally, a human will start to feel electrical current at a level around 1-5 ma. These currents are generally perceptible but produce little if any actual pain. At currents above 10 ma, currents in the body start to cause muscle contractions, which can lead to the inability to let go of a "hot" contact, once it is touched. This inability to let go allows the current to flow for longer times and can lead to serious injury or death. Currents above 100 ma can interfere with heart and/or respiratory function.

Resistance in the current path is what keeps current from reaching a dangerously high level. The resistance of the interior cells in the human body (largely composed of water) is generally quite low. The dry surface of human skin is much more insulating, typically adding a resistance of the order $10^5 \, \Omega$ to the current path. This is why it's critically important that your hands be completely dry when working around electronics equipment. Wet skin allows much more current to flow into your body from any power source than would be the case if your skin were dry.

The voltage levels of less than 20 V mentioned above come from the DC power supplies, function generators, and DAQ board input/output connections you must unavoidably be exposed to. For safety, make sure no other power sources, such as 110 or 220 V line power, are exposed at or near your lab bench in such a way that you might accidentally contact them. If you even suspect that such a power source might be exposed, report it immediately to your lab instructor or technician and keep everyone away until the situation is corrected.

When you are assembling a circuit, disconnect all voltage sources until you are ready to perform a test. Always wear shoes and avoid touching pipes or other metal fixtures while working in the electronics laboratory.

If someone else in your laboratory is the victim of electrical shock, the first thing to do is to shut off power to the conductor the victim is touching. If this is not possible, you need to break the contact of the victim with the conductor using something insulating like wood, rope, leather, or cloth. **Don't touch the victim yourself – you may become part of the conducting path.** Severe electrical shock often results in loss of consciousness and/or cessation of breathing. If you have forgotten or never learned artificial respiration/CPR techniques, now is a great time to review or get instruction in them. First aid for the burns often associated with shock may also be required.

Other laboratory hazards include those not directly related to electricity. Most likely, you will be building your circuits on a prototyping board that eliminates the need for soldering connections. If not, you need to treat the tip of the soldering iron like the very hot object it is.

Equipment hazards

The digital acquisition (DAQ) boards inside your computer in a laboratory like this one are delicate and expensive. Your introductory electronics laboratory probably has a protective interface placed between your circuit and the DAQ board to prevent excessive voltages being applied to board components. Do not attempt to defeat the purpose of this interface. If you suspect it is malfunctioning, do not try to wire around it. Instead, report the problem to your lab instructor or technician. Direct connections to DAQ board terminals should be made only by very experienced users.

Many of the projects in this workbook utilize operational amplifiers. An easy and lethal (to the op-amp, anyway) mistake to make is connecting DC power supply voltages to the DC supply leads of the op-amp with the wrong (reversed) polarity. If you do this, the op-amp usually expires immediately.

Chapter M1: Signal Generation

Introduction

Most of our understanding of how electronic circuits operate is based on analyzing these circuits in order to determine their response to some input waveform. The job of the analysis is to predict the voltage waveform, $v_o(t)$, that appears at some output node in the circuit when the circuit is excited by a given input waveform, $v_i(t)$. On the basis of these inputs and outputs, we determine properties of the circuit such as impulse- and step responses, impedances and admittances, transfer functions, and the like. Predicting such circuit behavior with analytical methods is the subject of a wide range of courses in the typical electrical and/or computer engineering curriculum. The laboratory part of the curriculum, however, is devoted to testing real circuits in order to compare measured outputs to those predicted by the analytical techniques.

Thus, it should not be surprising that signal generators are key equipment items in any electronics laboratory. They provide a wide range of standard waveforms such as sine waves, square waves, triangle and sawtooth waves, and periodic exponential waves, whose amplitudes, frequencies, dc offsets, and duty cycles can be adjusted easily. Many modern function generators let you program them to output custom waveforms, which you can define by means of mathematical functions or lookup tables. Both *Virtual Bench* and *LabVIEW* make it easy to generate voltage waveforms at the output terminals of your DAQ board.

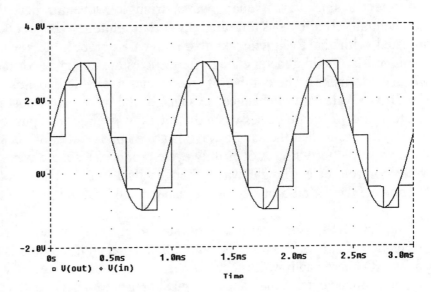

Fig. M1-1. The relationship between the programmed waveform, $v_p(t)$, and the output, $v_o(t)$, from a digital function generator.

M1.1. Digital Signal Synthesis

Most modern function generators are digital rather than analog. As such, they compute what their output should be at the time, t, from an operator-supplied mathematical function or table. The function generator converts the computed output value, $v_o(t)$, from a digital number inside the computer to an analog voltage that appears at the DAQ board output terminal. While the digital-to-analog conversion (DAC) is taking place, the next value to be output must be calculated or looked up in the table. The time this process takes imposes an upper limit on the update rate of the analog output. Between successive updates, the analog output voltage is constant. Consequently, the analog output is an approximation of the desired waveform, and is related to it as shown in Fig. M1-1. Clearly, the higher the update rate, the more closely the DAQ analog output, $v_o(t)$, resembles the continuous time function, $v_p(t)$, that the operator programmed. Also, if you want to produce accurate signals such as high-frequency sine waves, a very high update rate is required.

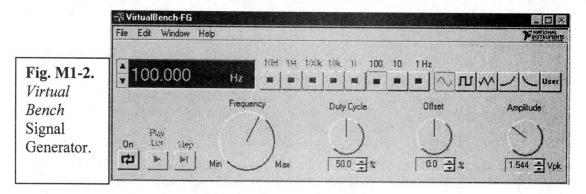

Fig. M1-2. *Virtual Bench* Signal Generator.

M1.2. The *Virtual Bench* Function Generator

Figure M1-2 shows the front panel of the *Virtual Bench* Function Generator. You can select a sine wave, square wave, triangle wave, or periodic rising or falling exponential waves by pushing the appropriate button. The frequency of this wave is adjusted with the **Frequency** knob. You will not be able to adjust the frequency continuously, but in general enough frequencies are available for the measurements you will need to make. The number of frequencies available depends on your DAQ board. The **Duty Cycle** control changes the shape of the wave. It is most useful if you want to create a train of square pulses when the square wave button is pushed. In this case, 10% duty cycle means that the square wave is at its upper value for 10% of the period and at its lower value for 90%. A 50% duty cycle produces a square wave. Adjusting the duty cycle control for a triangle wave changes its form to a sawtooth, by adjusting the fractions of the period occupied by the rising and the falling parts of the waveform.

The DC **Offset** control is calibrated in percent instead of volts. The reason for this choice of units arises from the output limits that all DAQ boards have. Depending on your DAQ board and how it is configured, it can output a range of voltages, but it cannot output a voltage outside that range. Typical ranges for DAQ boards are –10V to +10v, -5V to +5v, and 0 to 10v. Most DAQ boards configured for use in an electronics laboratory are configured for a bipolar voltage range such as –10V to +10v. For such a configuration, setting the DC offset control to 10% produces a dc offset of 10% of the maximum voltage magnitude, or +1 volt. Your lab personnel can tell you how your DAQ boards are configured, or you can measure the dc voltage levels associated with different percent settings of the DC offset control. The **Amplitude** control sets the peak amplitude of the waveform. The amplitude is measured from the voltage level of the DC offset to the maximum (or the minimum) voltage level in the waveform. You can operate all of these controls, except the one for **Frequency,** by turning the knobs with the mouse, or by typing the numerical values you want in the boxes. You can only adjust the output frequency by turning the knob.

The limited output range of your DAQ means that some combinations of DC offset and amplitude settings will take the maximum or the minimum value of your waveform out of the allowed output voltage range. If this happens, the warning, "CLIP," appears in the digital indicator for frequency on the front panel. During periods when the programmed waveform, $v_p(t)$, is outside the allowed range, the real output, $v_o(t)$, will remain at the voltage limit nearest to $v_p(t)$.

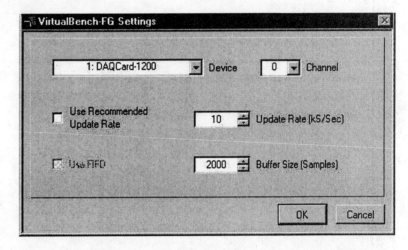

Fig. M1-3. The Settings box for the *Virtual Bench* Function Generator.

You can make some adjustments to your *Virtual Bench* Function Generator by selecting **Edit-> Settings** from the menu. You will then see a box like Fig. M1-3. The **Device** box will show any compatible DAQ boards in your system by name. The **Channel** box lets you select which analog output channel of the selected device will carry the output voltage, $v_p(t)$. You can use the update rate recommended for your device by checking the **Use Recommended Update Rate** box, or you can select your own update rate. The maximum update rate you can use successfully is determined by the capabilities of your DAQ board, your computer, and whether or not your computer is running any programs in background in addition to *Virtual Bench*. If you select an update rate that *Virtual Bench* knows is too high for your device, it will warn you immediately. However, selecting too high an update rate can cause *Virtual Bench* to run very slowly, or even lock up. If this happens, consider reducing the update rate. Your DAQ board may have a FIFO (first-in-first-out) register to allow faster update rates than would otherwise be available. If the **Use FIFO** option is available in your settings box, it is generally good practice to select it by checking the box.

Prep Exercise M1-1. Suppose the allowed input voltage range on your DAQ board is from -10 to +10 volts. Make a rough sketch of about 2 cycles of the output waveform, $v_p(t)$, for each of the following situations:

Sine wave: Amplitude = 1V, Offset = 10%, Frequency = 100 Hz, Update rate = 800 Hz

Sine wave: Amplitude = 5V, Offset = 70%, Frequency = 100 Hz, Update rate = 20 KHz

Sine wave: Amplitude = 1V, Offset = 10%, Frequency = 100 Hz, Update rate = 20 KHz

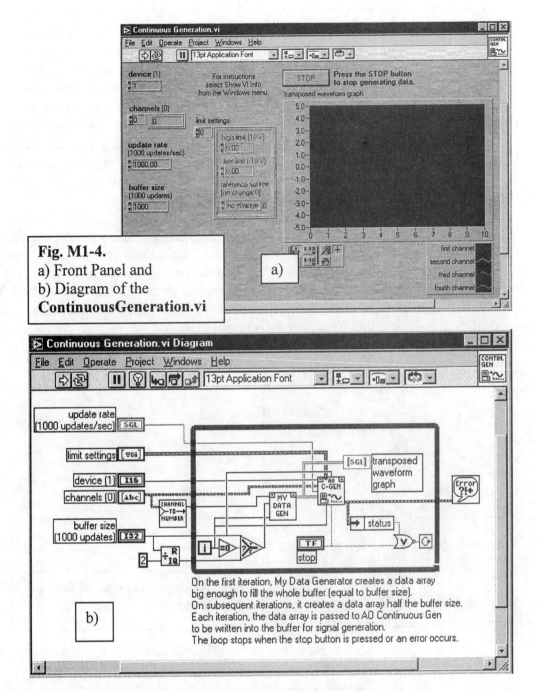

Fig. M1-4.
a) Front Panel and
b) Diagram of the
ContinuousGeneration.vi

M1.3. Signal generation using *LabView*

LabVIEW provides some example VI's in the **anlogout.llb,** which you can use for routine signal generation tasks with very little modification. You can build a useful function generator VI for your experiments starting from the **ContinuousGeneration.vi** located in the Examples->daq->anlogout directory. Its front panel and diagram are shown in Fig. M1-4. You can get a detailed description of the operation of this VI (or of any other *LabVIEW* VI in the Examples libraries) by selecting **Show VI Info** from the **Windows** menu once you have opened the VI in *LabView*. The **ContinuousGeneration.vi** uses the **My Data Generator** sub-VI to fill a buffer with a series of updates that describe the waveform to be output. The contents of this buffer are continuously sent to the DAQ board analog outputs with one new sample sent to each output channel at the **update rate** you select. One new value is sent from the buffer to the DAQ board analog output every (update rate)$^{-1}$ seconds. Note that the **buffer size** is set by a front panel control on this VI. The buffer size must be chosen so that the buffer is neither completely emptied as data is read out of it to the analog output, nor completely filled by new data from the *LabVIEW* VI, causing old data to be overwritten before it is read. The **device** control selects which DAQ board will receive the generated waveform data. The **channels** control is a list of channel numbers to be used as analog outputs, with channel numbers separated by commas, such as "1,0." Many DAQ boards have two or more available analog output channels.

To make a function generator VI useful for laboratory work, you can use this VI as a base, and make a few modifications. The type of waveform you get is determined within the **My Data Generator sub-vi.** To enable you to select from a variety of waveforms with adjustable amplitudes and frequencies, you will need to replace **My Data Generator** with a new sub-VI, named **MyFG**, as well as make a few modifications to the rest of the **ContinuousGeneration.vi**.

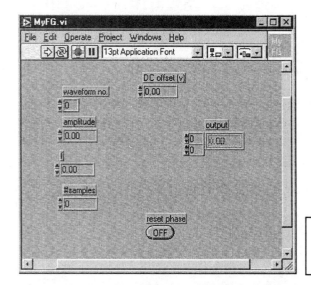

Fig. M1-5. Front panel of the **MyFG** sub-VI.

Figure M1-5 shows the front panel of the **MyFG** sub-VI. The **waveform no.** control selects the type of waveform from the following list:
        0 – Gaussian white noise
        1 – sine wave
        2 – square wave with a 50% duty cycle
        3 – triangle wave with a 50% duty cycle.

The **amplitude** control sets the peak amplitude of the sine, square, or triangle wave or the standard deviation of the Gaussian white noise in volts. The **DC offset** control provides a selected dc offset to the output of one of the two analog output channels. The **reset phase** control provides for continuous generation of the periodic waveforms (sine, square, and triangle) by allowing the output to start each new data generation at the point where it stopped the previous one. For more information about this feature, pull up **Online Help** for the **Sine wave.vi** in *LabView*. The *LabVIEW* VI's that generate the sine, square, and triangle waves require a dimensionless frequency input in units of cycles per sample. This means that, if you're using an update rate of 20KHz and you want to generate a sine wave with a real frequency of 1 KHz, the value of f that should be supplied to the **Sine wave.vi** is 0.05, since there would be 20 samples per cycle or 0.05 cycles per sample. In general the dimensionless frequency is given by,

$$f = \frac{f(Hz)}{update\_rate}. \tag{M1-1}$$

Figure M1-6 shows the diagram of the **MyFG** sub-VI. It contains a Case Structure with four cases corresponding to the numbers selected by the waveform control. Each case is shown separately in Fig. M1-6. In each case, one channel of the analog output contains the generated waveform while the other contains the same waveform plus a DC offset. You can consult **OnLine Help** in *LabView*, or the *LabVIEW VI Reference Manual* for more detailed information on each of the wave generation VI's if you wish.

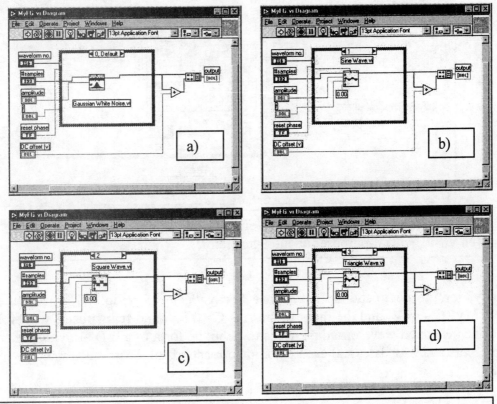

Fig. M1-6. Diagram of the **MyFG** sub-VI for each of the four cases: noise, sine wave, square wave and triangle wave.

The **MyFG** sub-VI then gets wired into the **Continuous Generation.vi** as shown in Fig. M1-7. You add the following front panel controls and indicators to the main VI:
1. **Frequency:** This is the frequency of any periodic waveform, in Hz. Its value, and the value from the **update rate** control are used to calculate the dimensionless f supplied to the **MyFG** sub-VI inside the While Loop. Since it is often useful to know how many samples per cycle will be generated, this value is calculated and brought to a front panel indicator.

2. **DC offset:** This value must be supplied from the front panel of the main VI to the **MyFG** sub-VI.
3. **Amplitude**: This value must be supplied from the front panel of the main VI to the **MyFG** sub-VI.
4. **Waveform:** This value is passed to the **MyFG** sub-VI, where it selects the case in Fig. M1-6 that corresponds to the desired waveform.

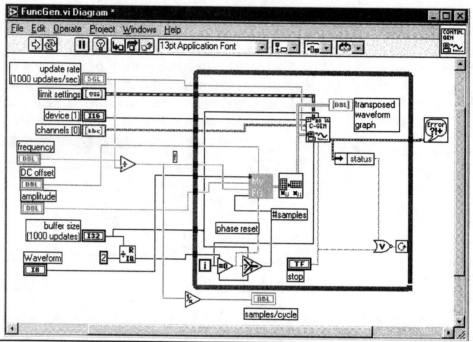

Fig. M1-7. Diagram of the **Continuous Generation.vi** as modified to make the **FuncGen.vi**, a general-purpose function generator.

Finally, don't forget to add the **Transpose Array.vi** shown in Fig. M1-7 between the output of **MyFG** sub-VI and the input to the **AO C-Gen.vi** and **transposed waveform graph**. The generated waveform data needs to be in the form of a 2-D array where each row represents a new update, and each column corresponds to a particular analog output channel.

Figure M1-8 shows how the front panel looks when generating a sine wave with a non-zero DC offset applied to one of the channels. Note that the first analog output channel listed in the **channels** control appears as the first plot on the transposed waveform graph. The order of the plots on the graph corresponds to the order of the output channels in the list.

```
Lab Skill Exercise M1-1.  Build the FuncGen.vi described in
this section, including the MyFG sub-VI.  Without
connecting its outputs to anything, run your VI and observe
the effects of the different front panel controls on the
waveforms plotted on the graph.  Set up the controls to
produce a sine wave of 1.0 volts amplitude, 1.0 volts DC
```

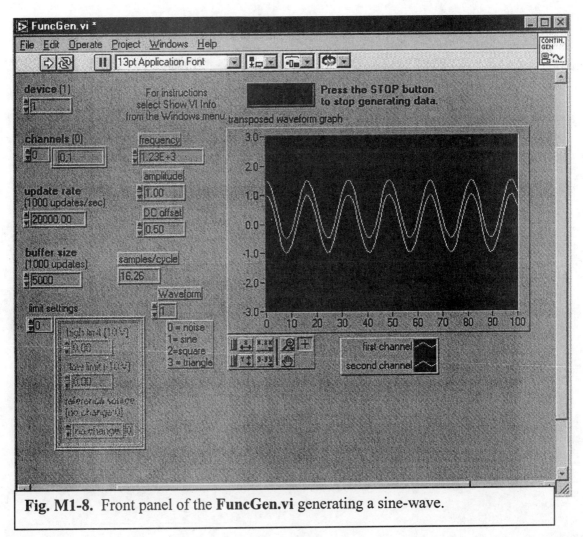

Fig. M1-8. Front panel of the **FuncGen.vi** generating a sine-wave.

offset, at the highest frequency for which you can obtain 15 samples per cycle. First, raise the **update rate** as high as your DAQ board will allow. Then increase the frequency until the **samples/cycle** indicator drops to about 15. Then push the stop button and adjust the scales on the graph so that approximately 3 cycles of the output on channels 0 and 1 are shown. Print out a copy of this front panel, and attach it in the space provided.

Save your **FuncGen.vi** for future lab work.

Chapter M1 *LabVIEW* Reference Table:			
Topic	Source		
	LabVIEW Example VI's (On-line)	*Learning with* **LabView**	*LabVIEW for Everyone*
Basics	All Fundamentals categories	Chs 1 & 2	Chs. 2-5
Structures	Fundamentals-> Structures, Graphs	Ch. 5	Ch. 6
Arrays and Graphs	Fundamentals-> Arrays, Graphs	Chs. 6 & 7	Chs. 7 & 8
Waveform Generation	DAQ	Ch. 8	Chs. 10 & 11
Use of Sub-VI's		Ch. 4	Chs. 3 & 4

Chapter M2. Waveform Capture

Introduction

The electronic devices and circuits of interest to us respond with such speeds that no human could actually "observe" their outputs in real time. This is why the waveforms characteristic of electronic circuits need to be "captured" so we can view them. With older oscilloscopes, the waveforms had to be excited repeatedly, and the display of the response triggered at the same point during each of these repetitions, in order to produce a stable display for human viewing. To illustrate this process, consider the measurement of the output response of the series RC circuit in Fig. M2-1a) to a one-volt step function.

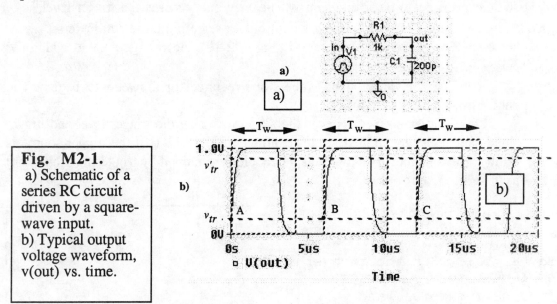

Fig. M2-1.
a) Schematic of a series RC circuit driven by a square-wave input.
b) Typical output voltage waveform, v(out) vs. time.

The function generator is set to output a square-wave which goes between the levels of zero and one volt, at a frequency which is low enough so that the output voltage finishes its exponential excursions in both directions before the input switches to the opposite level. The resulting output waveform, v(out), is shown in Fig. M2-1b).

• **Display mechanism.** For many decades, the standard laboratory instrument for capturing and displaying waveforms was the analog oscilloscope. In an analog oscilloscope, each voltage to be displayed is continuously converted into a vertical deflection of a spot on the screen, while the spot is swept across the screen at a rate determined by the **sweep-time/cm** control of the scope. Thus, if the screen is 10 cm wide, and the sweep-time/cm control is set to 0.1 ms/cm, the display is a 1 ms "window" showing the input and output voltages vs. time. The time, T_w, represents this window width in Fig. M2-1b). On a typical cathode-ray tube (CRT) screen, this display will only last as long as the bright spot on the screen persists, which is generally too short a time to be useful in taking measurements.

- **Triggering.** In order to get a stable display that lasts long enough for measurement purposes, two things have to happen:
  1) The sweep of the traces across the screen must be repeated often enough so the human eye sees a continuous image of the window, and
  2) This repeated sweep must begin at an equivalent point on the waveform each time the sweep occurs.

The points A, B, and C in Fig. M2-1b) represent three equivalent points in the output waveform. If three successive screen sweeps were triggered at these points, the appearance of the oscilloscope screen during each sweep would look the same, as shown by the three successive windows (dashed squares) in Fig. M2-1b).

- **Triggering Parameters.** In the example of Fig. M2-1b), the triggering circuitry in the oscilloscope is set to begin a sweep whenever v(out) crosses the trigger level, v_{tr}, with a positive slope. The oscilloscope has controls for selecting these parameters:
  - **Trigger level.** This is v_{tr} in Fig. M2-1b), the voltage level at which the sweep is begun.
  - **Trigger slope.** This is the slope required for a sweep to begin. The choice is positive or negative.
  - **Trigger source.** This is the signal to which the trigger level and trigger slope parameters are applied. In the example of Fig. M2-1b), the trigger source is v(out). However, a very similar display could be obtained by triggering from v(in) as the trigger source.

Prep Exercise M2-1.
Suppose your oscilloscope is displaying the waveform in Fig. M2-1b), with the trigger level set to v_{tr}', trigger slope set positive, and trigger source set to v(out). Sketch here how the display would look. Assume T_w is approximately the same as it appears in the Figure.

Prep Exercise M2-2.
Suppose your oscilloscope is displaying the waveform in Fig. M2-1b), with the trigger level set to v_{tr}', trigger slope set negative, and trigger source set to v(in). Sketch here how the display would look. Assume T_w is approximately the same as it appears in the Figure.

- **Storage of the displayed waveform.** In the above example, the oscilloscope screen was refreshed with the same image every time a new sweep was triggered. In order to store these waveforms for a lab report or for later processing, you would need to make a drawing or a photograph of the oscilloscope screen. Since so much of today's information processing and report preparation takes place within personal computers, it seems reasonable to find a way to store the waveforms themselves in the memory of the computer. You could then pass them to other software which would graph them on the computer screen, perform mathematical operations on them, or produce a hard copy of the graph on the screen. This is the approach used by most modern digital oscilloscopes.

One big advantage of being able to store waveforms is that they only have to be captured once instead of repeatedly. Other advantages of this approach will become apparent in the following sections.

M2.1 Capturing Waveforms : Digital Representation

M2.1.1 Basics of Sampling

Your computer does not acquire the entire time-continuous waveform it receives, but is only taking samples of the signal at regular time intervals. Thus, in the computer, a time-dependent voltage becomes a list of voltage values (regularly spaced in time) called a **time series**. The time series is, of course, only an approximation to the waveform, $v(t)$, which has been sampled. It is worthwhile to gain some understanding of the conditions under which this approximation is a good one. To do this we will digress into some theory here.

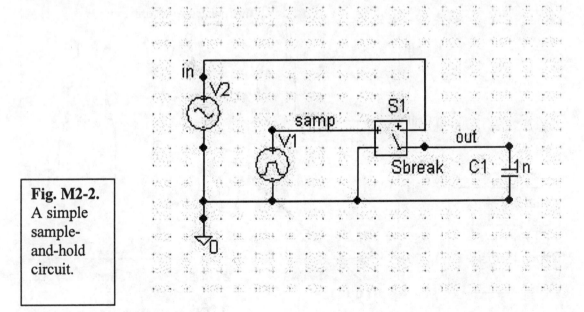

Fig. M2-2. A simple sample-and-hold circuit.

Consider the circuit shown in Fig. M2-2. It is a simple example of a "sample-and-hold" circuit. The device labeled "S-Break" in Fig. M2-2 is a voltage-controlled switch that closes whenever it sees a positive input from the source, V1. The inputs and output

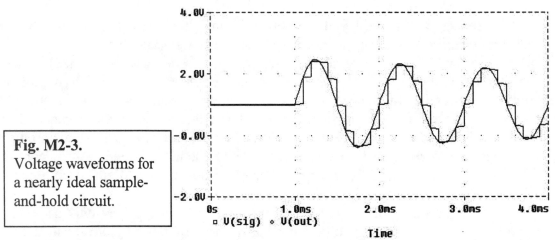

Fig. M2-3. Voltage waveforms for a nearly ideal sample-and-hold circuit.

of this circuit are shown in Fig. M2-3 for the case of a low-frequency, damped sine wave input and a reasonably high sampling rate. The output of this circuit is a stair-step approximation to the input. The output values in the flat part of the stairs are the values of the input voltage when the switch was turned off at the end of each sampling pulse. This circuit gets its name from the sequence of operations it performs:

- A sample of the input voltage is placed on the capacitor each time the voltage-controlled switch is closed.
- The capacitor holds the sampled signal until the next sample is taken. During the time interval between samples, the computer has time to do an analog-to-digital conversion and to store the result as one value in the time series that now represents the signal in the computer.

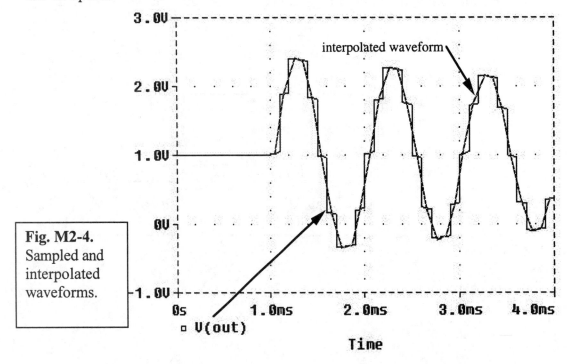

Fig. M2-4. Sampled and interpolated waveforms.

There is another way of looking at what the computer actually "knows" about this signal once the sampling process and the analog-to-digital conversion are finished. Figure M2-4 shows the output of the sample and hold circuit. In the approximate center of the sampling interval, a sample point has been drawn at the sampled voltage value. A reasonable way to approximate the signal between sample points is linear interpolation. The result of connecting the sample points with straight lines is also shown in the figure. The straight lines connect points that lie approximately at the middle of each sampling period.

This linear interpolation is the way *Virtual Bench* Oscilloscope and most other digital oscilloscopes draw the signals they receive to a graphical user interface (GUI): the signal in the interval between two sample points will always be drawn as a straight line. Clearly, this linear interpolation will not be a good approximation to the input unless samples are taken often enough to follow the variations in the input. Figure M2-5 shows the effect of increasing the oscillation frequency of the input by a factor of eight. The interpolation is now a pretty poor representation of the input signal. The interpolation has pointed tops and bottoms, in contrast to the input sine wave.

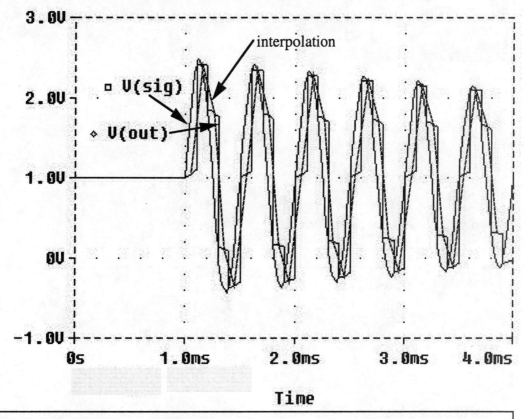

Fig. M2-5. Input [V(sig)] and output [V(out)] of the sample-and-hold circuit, along with a straight-line interpolation joining points that lie approximately at the middle of the sampling period.

A good intuitive way to think of this high-frequency limit is to visualize how many dots are needed to make a "good" drawing by connecting the dots with straight lines. The mathematicians have come up with methods for getting along with even fewer samples, by using curves considerably more complex than straight lines to connect the dots. Since most digital oscilloscopes do not use these more complex interpolations to reconstruct waveforms from time series, we will not discuss them further here.

Prep Exercise M2-3. The Whittaker-Shannon Sampling Theorem states that waveforms containing frequencies up to B can be exactly recovered by sampling at a frequency $\geq$ 2B. Sketch three cycles of a sine wave here. This will represent a sine wave of frequency B. Now draw on that curve 12 equally spaced dots which would represent samples taken at the frequency 4B (twice the minimum required sampling frequency, according to this theorem). Connect these dots with straight lines, and comment on how good an approximation to the original sine wave you have constructed.

Sketch:

Comments:

M2.1.2 *Virtual Bench* Oscilloscope

It is not at all necessary that the controls and display for a digital signal acquisition system look like those of the analog oscilloscope that most older engineers are accustomed to. However, for those having some previous experience with oscilloscopes (a large fraction of the engineering population), the transition to digital signal acquisition may be somewhat easier if the graphical user interface (GUI) has the more recognizable look and feel of a conventional oscilloscope. The *Virtual Bench* Oscilloscope is just such a Virtual Instrument.

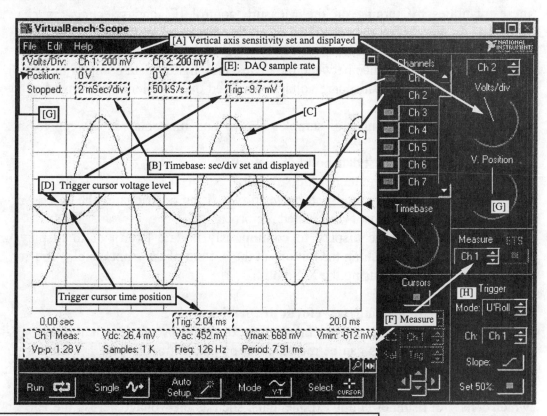

Fig. M2-6. The *Virtual Bench* Oscilloscope front panel.

The front panel for the *VB* Oscilloscope (using typical default panel settings) is shown in Fig. M2-6. The scope will display signals for those input channels whose buttons are pressed [B]. You can set the vertical axis sensitivity [A], and the vertical position [G] independently for each of those input channels. To set the vertical sensitivity for Channel N, push the arrow buttons immediately above the Volts/div. knob until the correct channel number appears between the arrows. Then make the sensitivity and position settings.

The Auto-Set-up button at the bottom of the screen lets the computer try to make the appropriate setting for the display based on the waveforms the DAQ board has just acquired. This is a good way to start looking at a waveform, but it may not provide exactly what you want.

In the Trigger control area [H], you can set trigger mode, slope (rising or falling) and select which channel you want to use as the trigger source. You set the horizontal and vertical position of the trigger cursor [D] by dragging it on the screen. Since this scope works with digitally stored signals, it can display the part of the signal that occurs before the triggering event.

The *Virtual Bench* Oscilloscope can also make some calculations on the signal coming in on a selected channel, which are displayed below the graph [F]. You display these quantities by turning on the Measure control [F] and selecting the channel to measure. In Fig. M2-6, the calculated quantities displayed are:
Vdc : the average dc level
Vp-p : the difference between the lowest and highest data point on the graph.
Vac : the root-mean-square voltage, averaged over the length of time displayed on the graph.
Samples: the number of samples shown and interpolated between to make the graph.
Vmax : the highest data point displayed
Vmin : the lowest data point displayed.
Freq and Period : The time between two rising crossings of the Vdc level is the period and the frequency is its calculated reciprocal.

You can cause the display to continuously update [Run], or to display only a single run [Single] by pushing the buttons at the bottom of the front panel. You can scroll the display left and right using the X-Scroll bar.

M2.1.3. Measuring waveform properties

Electrical engineers have defined many different properties that describe signals. Many of these properties can be measured using calculations on the acquired waveform. Comparing the amplitude and phase of two sine waves is one obvious example, which is presented in Chapter M4. Another important property is the time-constant of signals which consist of rising or falling exponentials, such as the voltage across the capacitor on Fig. M2-1.

A. Voltage and time differences.

The scope screen shown in Fig. M2-6 is displaying two sine waves that have the same frequency, different amplitudes, and are phase-shifted with respect to each other. Comparing the amplitudes of these two sine waves is easy: Just use the Measure control on each channel and compare the two values of Vp-p. The value for Vp-p is twice the amplitude.

You can also use the Cursors on the *Virtual Bench* Oscilloscope to determine voltage and time differences. Figure M2-7 shows the same waveforms as Fig. M2-6, with the two available cursors, C1 set on the maximum and C2 set on the minimum, of the Channel 1 waveform. The cursor channel indicators identify the displayed waveforms to which each cursor is attached.

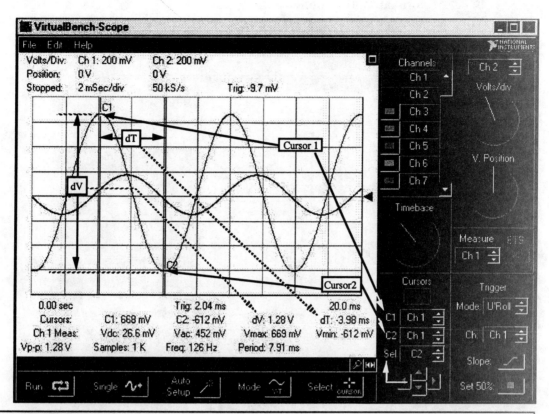

Fig. M2-7. Use of the Cursors in the *Virtual Bench* Oscilloscope to determine voltage and time differences.

When the cursors are on, a new display line appears below the scope screen, which shows the voltages at each cursor location [C1 and C2], the difference in voltage, dV, between the two cursor levels, and the time difference, dT, between the two cursor locations.

Typically, you move the cursors around the display by dragging them with the mouse. As an alternative, the cursor control labeled **Sel** lets you select which of the two displayed cursors you want to move by clicking on the arrows immediately below the **Sel** control.

B. Time-constants of exponentials.

You can measure the time-constant of a simple RC network using the *Virtual Bench* Oscilloscope a couple of different ways. They are summarized here.

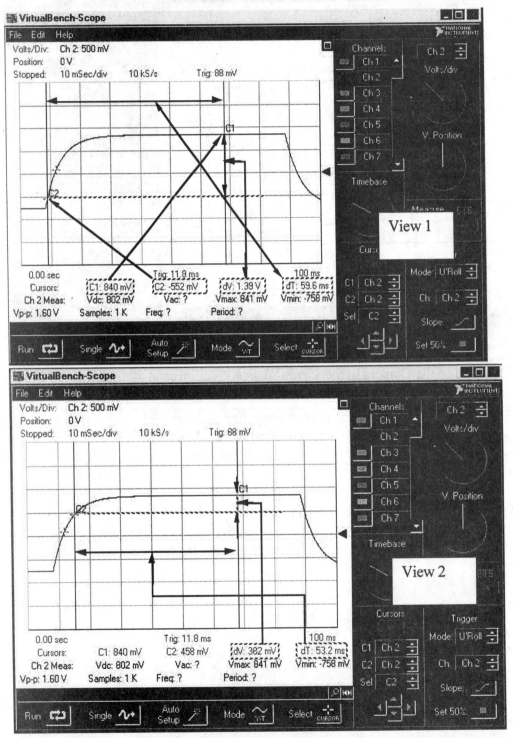

Fig. M2-8. The same rising exponential with two different Cursor placements in view 1) and view 2) for measuring the time constant.

Method 1. This method works well when you can see the steady-state value the exponential rise or fall is approaching. Suppose the circuit in Fig. M2-1 were driven with a low-frequency square wave. In this example, the square wave drive is slow enough so that V(out) reaches a steady-state value each time the square wave drive switches sign. Figure M2-8 shows two views of a rising exponential on Ch2, with the cursors positioned differently in the two views.

In both views, cursor C1 is in the same position, at the "top" of the rising exponential where the voltage has reached a steady state value. In view 1), cursor C2 is 1.39V below this steady-state value. In view 2), cursor C2 has been moved and is now 0.382 V below the steady-state value, at a time later than in view 1. The voltage *relative to the steady-state value, marked by cursor C1,* obeys the equation for a falling exponential and relates the measurement values in views 1 and 2 of Fig. M2-8 by

$$dV_1 \exp\left\{-\frac{dT_1 - dT_2}{\tau}\right\} = dV_1 \exp\left\{-\frac{\Delta t}{\tau}\right\} = dV_2, \tag{M2-1}$$

where τ is the desired time-constant and the subscripts 1 and 2 refer to the number of the view in Fig. M2-8. Since the measurement data from Fig. M2-8 provide numerical values for everything except the time-constant, solving for it by taking the natural log of both sides and rearranging terms gives

$$\tau = -\frac{\Delta t}{\ln\left\{dV_2/dV_1\right\}}. \tag{M2-2}$$

`_Prep Exercise M2-4:` Calculate the time-constant of the exponential in Fig. M2-8 using the above method.

Method 2. This method works when your measured signal doesn't clearly show the steady-state value of the exponential. Figure M2-9 shows two sets of cursor placements which can produce a measured time constant in this situation. In view 1,

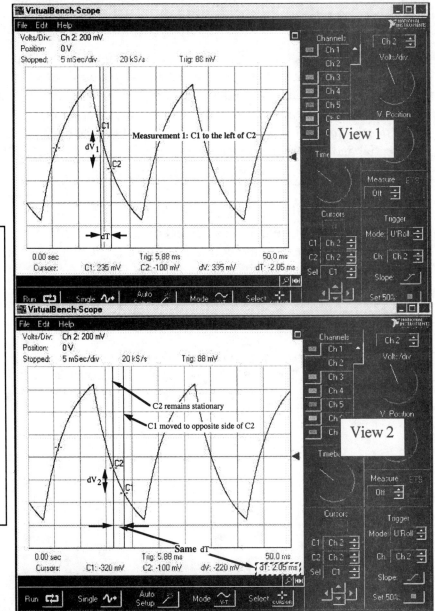

Fig. M2-9. The same exponential with two different Cursor placements in view 1) and view 2) for measuring the time constant, when the steady-state value cannot be seen.

cursor C1 is positioned a time interval dT =12.05 ms to the left of (earlier than) C2. In view 2, C1 has been moved an **equal** interval to the right of (later than) C2. Figure M2-10 illustrates how to find the time constant for this exponential even though you cannot see its final steady state value. In Fig. M2-10, we "pretend" we can extend the exponential to this final steady-state value. Since we can't see the steady-state value (the lower horizontal dashed line), we don't know the value of A in Fig. M2-10, and we must therefore eliminate it from our equations. If we define

$$x \equiv \exp\{-dT/\tau\},\tag{M2-3}$$

then Fig. M2-10 tells us the following:

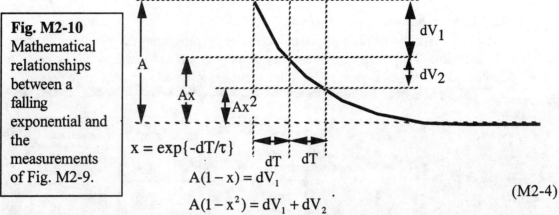

Fig. M2-10 Mathematical relationships between a falling exponential and the measurements of Fig. M2-9.

$$x = \exp\{-dT/\tau\}$$
$$A(1-x) = dV_1$$
$$A(1-x^2) = dV_1 + dV_2 \quad (M2\text{-}4)$$

Eliminating A from Eqs. (M2-4) by taking a ratio gives a single quadratic equation in x. One solution is $x = 1$. The other, more useful, solution is

$$x \equiv \exp\{-dT/\tau\} = \frac{dV_2}{dV_1}, \quad (M2\text{-}5)$$

so that the time constant in terms of dV_1, dV_2, and dT becomes

$$\tau = \frac{-dT}{\ln\left\{\dfrac{dV_2}{dV_1}\right\}}. \quad (M2\text{-}6)$$

In summary, you can calculate a time-constant from a set of cursor readings using Fig. M2-9 and Eq.(M2-6). *Just be careful that you use equal dT intervals for the two views.*

> **Prep Exercise M2-5:** Calculate the time-constant of the exponential in Fig. M2-9 using the above method. Show your work here.

M2.1.4 Free-standing oscilloscope

The *Virtual Bench* Oscilloscope was deliberately programmed to resemble many common freestanding oscilloscopes. These instruments come in too many varieties to be described in any detail here. Figure M2-11 shows an example of a typical front panel.

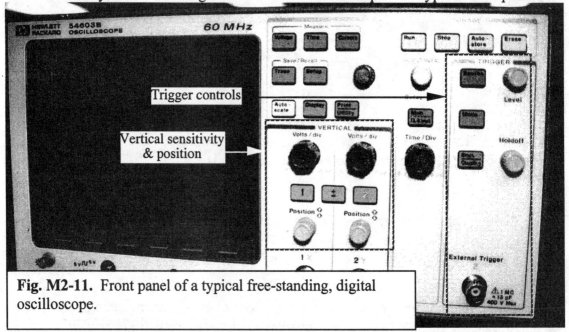

Fig. M2-11. Front panel of a typical free-standing, digital oscilloscope.

You will be making nearly all of the same adjustments and settings described above for the *Virtual Bench* Oscilloscope on your free-standing one, namely:
- vertical sensitivity in volts/division,
- sweep-time / cm,
- trigger type, source, etc.,
- vertical and horizontal positions of the display on the screen,
- cursor positions and readings (digital scopes only)

It is important to remember that these essential components of oscilloscope operation are common to nearly all instruments.

If your free-standing scope is an analog instead of digital, it will display the continuous waveform on its screen instead of the connected series of dots described here for digital scopes. Proper triggering will be especially important to get a stable display on the screen, and features such as single-shot triggering and cursors will generally not be available. This may lead you to conclude that all analog scopes are laboratory dinosaurs and that digital ones are superior for all applications. However, the fastest oscilloscopes, those that measure microwave and higher frequency signals, are mostly analog because it takes too much time to do all this sampling and analog-to-digital conversion described above.

M2.2 Signal Acquisition in *LabView*

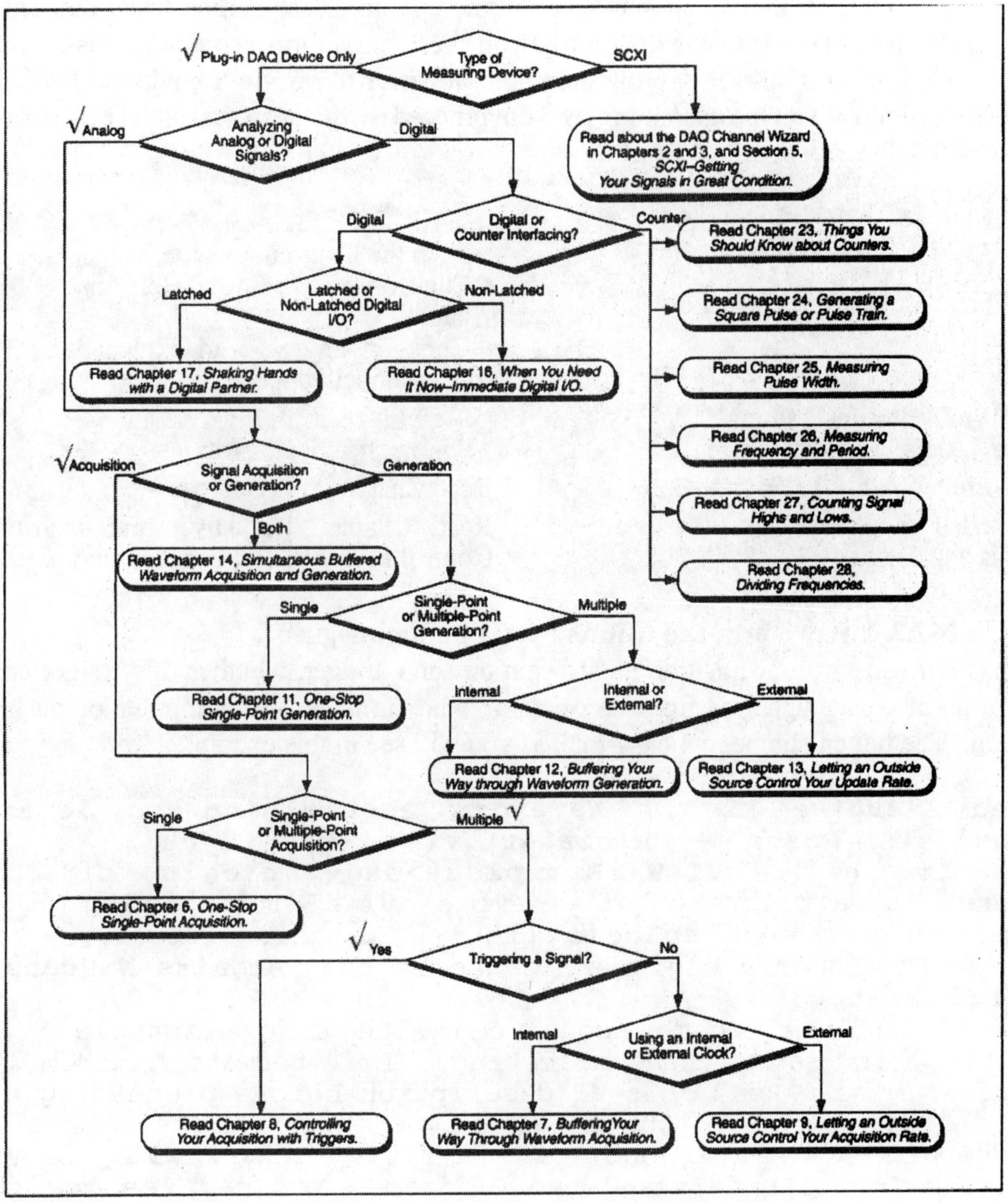

Fig. M2-12. Flowchart for the *LabVIEW* Data Acquisition Basics Manual. This chapter uses material in the check-marked path.

If you have *LabVIEW* installed on the computers in your laboratory, and if those computers are equipped with data acquisition (DAQ) boards, you can start acquiring signals from your lab bench about as easily as launching *Virtual Bench* or turning on a free-standing oscilloscope. The upside of using *LabVIEW* is the flexibility you have in processing your signals as you acquire them.

The Examples directory of *LabVIEW* contains virtual instruments (VI's) that you can use with minimal modification to do all of the waveform acquisition tasks that are the subject of this chapter. Our focus here is on these waveform acquisition tasks, rather than on the details of *LabVIEW* programming. However, if you're a novice *at LabVIEW*, you can learn a lot of programming as you proceed with these tasks. The following sources of information will be helpful:

Information Source	Procedure to follow
LabVIEW On-line Help	• Go to the Diagram on any vi • Click on Help->Show Help in the Menu Bar • Click on any vi : an explanation of its operation will appear
LabVIEW documentation: Data Acquisition Basics Manual	• See Flowchart in Fig. M2-12 • Follow the checkmarks
Learning with LabVIEW, Bishop	Read Chapter 8: Data Acquisition
LabVIEW for Everyone, Wells and Travis	Read Chapter 11: DAQ and Instrument Control in *LabVIEW*, beginning at p. 325

M2.2.1. Waveform acquisition with a human trigger

All waveform acquisition must begin on some trigger, whether this trigger comes from the incoming signal or from an operator who turns on the computer or pushes a button. The button can be real or virtual, as you will see in this example.

Prep Exercise M2-6. Waveform acquisition on demand: Acquire N Scans - Multistart.vi

In the **LabVIEW->Examples->daq->anlogin** directory there is a library of example VI's for analog input operations called **anlogin.llb**. In this library, the simplest VI for a beginner to use is the **Acquire N Scans - Multistart.vi**.

• You can become acquainted with this versatile VI by opening it, and then selecting VI information from the Windows menu. Read this VI description before proceeding.

Also, examine the Diagram of this VI. Don't be concerned if you're not yet familiar enough with *LabVIEW* programming to understand the Diagram. You can see how this VI works by using it.

• Use On-Line Help to get more information on any parts of the vi you don't understand.[See the Table above].

• The next step is to generate a signal to acquire. Design (on paper only, for the present) a simple series RC circuit like the one in Fig. M2-1. You will drive it with any handy square-wave source. The resistor and capacitor values you choose for this experiment don't matter very much.

• Enter your selected values for the resistor and the **capacitor** here (choose standard component values you can obtain in the lab -- you are going to build this circuit later):

```
Chosen standard values:

R = _____ (ohms)


C = _____ (farads).
```

• Calculate the RC time-constant τ for your circuit and enter the result here:

```



$\tau$ = _____ (s)
```

• In order to get an output voltage in Fig. M2-1 that looks like the one shown, both the upper and lower voltage levels of the square-wave input need to be several (> 3) time-constants long. For your value of τ, select a frequency for the square-wave that meets this criterion, and enter it here:

```



$f_{sqw}$ = _____ Hz.
```

• Suppose you want the VI to display about 2-3 cycles of the input and output to this circuit. How long a time, T_r, do this many cycles of your square-wave represent? Enter your answer here:

```



$T_r$ = _____ sec.
```

• To get an accurate representation of the signals to be acquired on the screen of this VI, **about** how may dots will be required for **each** signal to be acquired and plotted? This number will be entered at **number of scans to acquire,** one of the controls on this VI. Enter your estimate here:

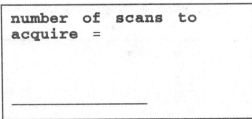

•Finally, in order to get that number of scans in the time period T_r, how fast does the **scan rate** (number of scans per second, another control on the VI) have to be? Enter your estimate here:

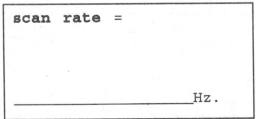

Figure M2-13 shows the front panel of the **Acquire N Scans – Multistart.vi**, with the modifications necessary to acquire the input and output signals from a circuit like that shown in Fig. M2-1a).

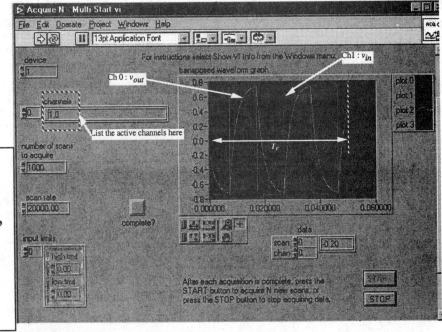

Fig. M2-13. The **Acquire N Scans – Multistart.vi,** modified to acquire two input channels.

Note that we have entered a list of the active channels and set the **number of scans** at 1000, with a **scan rate** of 20KHz. Thus, it shouldn't be surprising that the length of time, T_r, over which signal acquisition takes place is 0.05 sec. Pushing the **START** button at the lower right begins the waveform acquisition. The **STOP** button disables further waveform acquisitions.

The **STOP** button also enables other modifications to the VI while the recently acquired waveforms are still on the screen. Figure M2-14 shows two useful modifications:

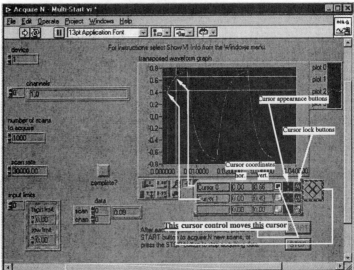

Fig. M2-14. Results of moving the data indicator with the arrow, and of creating a cursor display.

- a) dragging the data indicator to the left with the arrow tool, to make room for
- b) right-clicking on the transposed waveform graph and selecting Show->Cursor Display. This creates a cursor display which you can use to read the signals at selected times.

You can create as many cursors as you want for the display. In this example, three will be needed to measure the time-constant of the exponential output voltage. Clicking on one of the lower corners of the cursor display and dragging down creates one more cursor, as shown in Fig. M2-15. By clicking on the Cursor lock buttons, you can lock selected cursors to particular plots. The boxes show the horizontal and vertical coordinates of each cursor.

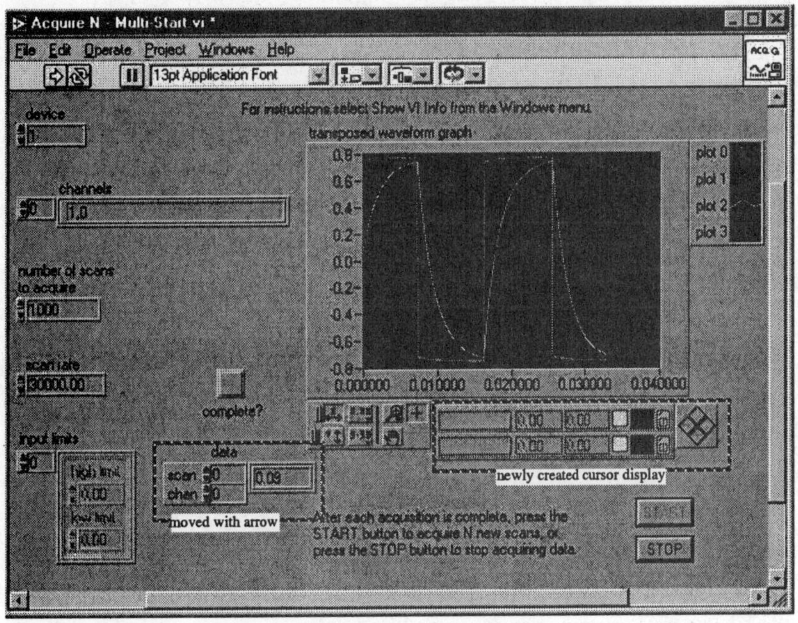

Fig. M2-15. Creating cursors for a waveform graph.

Cursor appearance buttons allow some modifications of what each cursor looks like. The cursor control moves all cursors for which you have pressed the square button to the left of the cursor appearance button.

In Fig. M2-16, the three cursors have all been locked to the exponential waveform, $v_{out}(t)$. The cursor display shows that the times for the tree cursors are equally spaced. You can use Method 2 from the preceding section (see Fig. M2-10) to determine the time constant of this exponential.

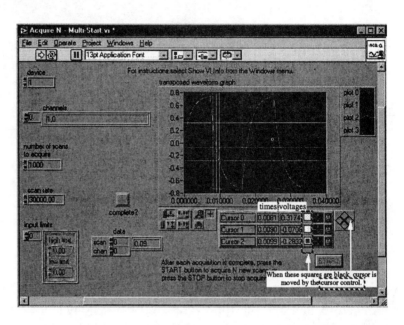

Fig. M2-16. Measurement of the time-constant of an exponential using three cursors, equally spaced in time.

Prep Exercise M2-7: Use Method 2 to calculate the time constant of the exponential waveform in Fig. M2.16.
Enter the values you read from the VI front panel here:

dT = _____ (sec)

dV_1 = _____ (volt)

Show your calculation of τ here:

τ = _____ (sec)

Lab Skill Exercise M2-1 :
- Build the circuit from Prep Exercises M2-6,7. Open the **Acquire N Scans - Multistart.vi** in *LabVIEW*, and make the modifications necessary to display 2-3 cycles of the input and output waveforms. Connect nodes (in) and (out) to the DAQ input channels you intend to use.
- Initiate the data acquisitions by pressing START. Create three cursors, and set them up equally spaced in time to obtain the time-constant of your output waveform, V_{out}.

- Print out a screen from your measurement that looks like Fig. M2-16, and place it here.

- Calculate the time constant of the exponential output waveform from your measurement.

$\tau =$ _____ sec.

- By what percent is your measured time-constant larger or smaller than your predicted one? Can the $\pm$ 10% variation in standard component values explain the difference?

M2.2.2. Waveform acquisition with the analog software trigger

To perform a triggered acquisition that behaves like an oscilloscope, you need to specify the triggering conditions based on some analog characteristic of one of the acquired waveforms. Figure M2-17 shows the front panel of the **Acquire N Scans Analog Software Trigger.vi** from the same library (LabVIEW->Examples->daq->anlogin).

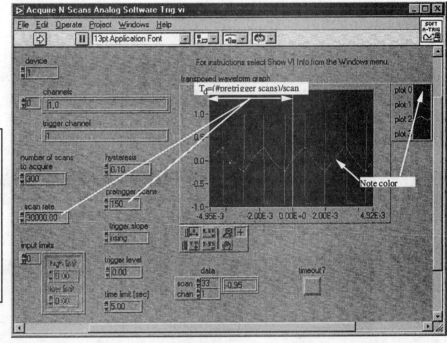

Fig. M2-17. The **Acquire N Scans Analog Software Trigger.vi** front panel.

The figure shows the front panel after a signal acquisition on Channels 1 and 0, with the VI set to trigger on Channel 1 (the exponential), rising slope, and a trigger level of 0 volts. This VI displays a cursor at the trigger point. Acquisition of data is continuous in this VI, with a display triggered when trigger conditions are met. In this example, the VI retains 150 **pretrigger scans** out of a total of 300 **scans to acquire**, so it is not surprising to find the trigger cursor located halfway across the screen. Note also that this VI measures time relative to the location of the trigger cursor, with pre-trigger times represented as negative.

Setting the pre-trigger scans control to zero and running the would result in the cursor being located off to the left of the display.

The **Acquire N Scans Analog Software Trigger.vi** closely resembles a virtual oscilloscope that scales its horizontal and vertical axes automatically. You can use it as a general "workhorse" VI for acquiring waveforms upon the occurrence of preset trigger conditions. You could measure the time-constants of the exponentials in Figs. M2-16,17 by creating cursors and performing the same calculations as in the preceding Lab Skill Exercise. But since one of the big advantages of writing your own virtual instrumentation in *LabVIEW* is the capability of automating your measurements, let's put some of the cursor manipulation and calculation into the VI. In the next Lab Skill Exercise, you will "customize" the **Acquire N Scans Analog Software Trigger.vi** to automate the measurement of exponential time constants.

Lab Skill Exercise M2-2: Using the Cursors and Graph Attributes to automate the measurement of time constants:

• Build the circuit from Prep Exercises M2-6,7. Open the **Acquire N Scans Analog Software Trigger.vi** in *LabVIEW* and make the modifications necessary to display 2-3 cycles of the input and output waveforms. Connect nodes (in) and (out) to the DAQ input channels you intend to use.

• Save your modified VI as **MY Acquire N Scans Analog Software Trigger.vi.** You will be making some extensive changes in this VI, and you don't want to modify the original one.

• Create an **Attribute Node** for **the transposed waveform graph:**
Go to the **Diagram**, right-click on **the transposed waveform graph** and select **Create Attribute Node.** The newly created node appears next to the graph to which it is attached, as shown in Fig. M2-19. If you right-click on an

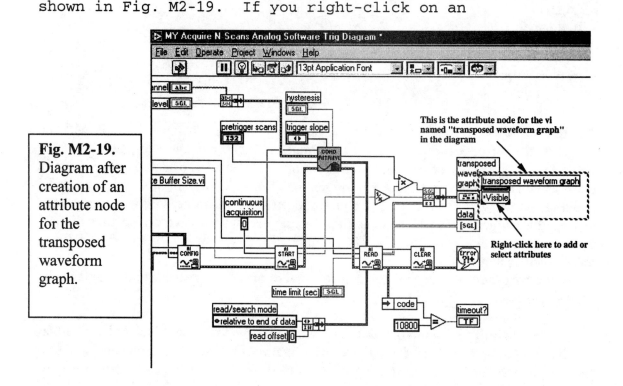

Fig. M2-19. Diagram after creation of an attribute node for the transposed waveform graph.

attribute [such as "Visible" in Fig. M2-19], a menu appears which gives you the chance to re-select the attribute that appears in the Attribute Node or add to the list of attributes.

• Modify the **Attribute Node** so that it displays on a Front Panel indicator the **Cursor Index** of Cursor 1:
Select the **Active Cursor** and **Cursor Index** attributes for operation with the Attribute Node. See Fig. M2-20 for how the result should look on the diagram. By default, the

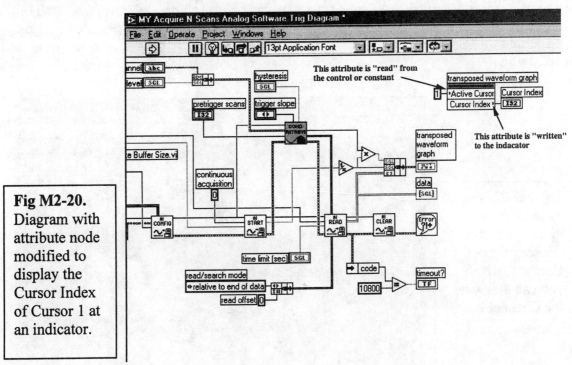

Fig M2-20. Diagram with attribute node modified to display the Cursor Index of Cursor 1 at an indicator.

Attributes you select appear as **Read** Attributes; that is, they read the values the attribute will then assume from controls or constants you wire into the diagram. **Read** Attributes have a little arrow at the left side, like the **Active Cursor** in Fig. M2-20. Note that in Fig. M2-20, this part of the attribute node is wired to a constant with a value of 1.

• By right-clicking on **Cursor Index** and selecting **Change to Write**, cause the little arrow to move to the right side of the attribute box for the Cursor Index. You have just changed the **Cursor Index** to a **Write Attribute**. This attribute can now supply a value that can be displayed on a Front Panel Indicator or otherwise used in any *LabVIEW* program you might wish to write.

• Wire the **Cursor Index Write Attribute** to a digital indicator so you can display its value. Your Diagram should now look like Fig. M2-20. Your Front Panel should look like Fig. M2-21.

• Your circuit should still be hooked up to the DAQ board in your computer. Make sure the drive voltage for the circuit is still on, and run your VI. After the input and output voltages for the circuit are displayed once again, lock Cursor 1 to the output voltage channel.[Click on the lock symbol and follow the menu that appears] Use the cursor tool shown in Fig M2-20 to drag the cursor right and left,

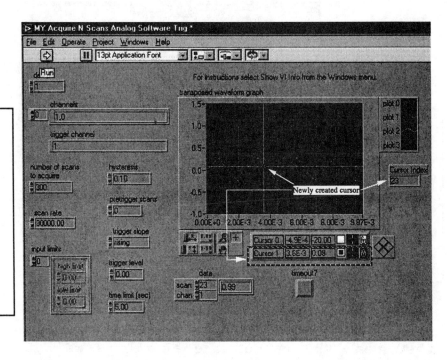

Fig. M2-21. Front Panel with an indicator of the Cursor Index of Cursor 1, which you can drag over the Channel 1 display.

and watch the value of the cursor index change, along with the x and y-coordinates of the cursor in the cursor box.

• Now you are ready to modify this VI further, so it will measure exponential time constants for you without requiring any manual calculations on your part.

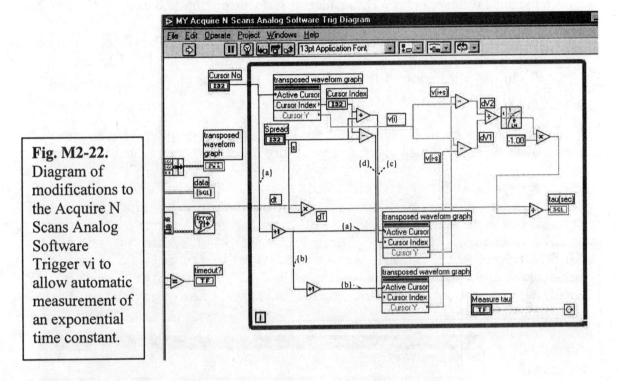

Fig. M2-22. Diagram of modifications to the Acquire N Scans Analog Software Trigger vi to allow automatic measurement of an exponential time constant.

a) Review Method 2 for exponential time constant calculation above. This is the method that requires three cursors equally spaced in time.
b) Examine the diagram shown in Fig. M2-22. until you understand its operation. First, read the explanation below:

This diagram is an addition to the **Acquire N Scans Analog Software Trigger.vi**. Its only direct connection to the original VI is via the data line labeled **dt** in Fig. M2-22 entering the **While Loop** on the left side. The constant which determined the **Active Cursor** has been replaced by a control, labeled **Cursor No.**, that will allow you to select which cursor will be at the middle of the group of three cursors used to calculate the time-constant. At the bottom of the While Loop you will see two data paths, {a} and {b}, which increment this **Cursor No.** twice. These paths then feed two more **Attribute Nodes** for the **transposed waveform graph**. These Attribute Nodes have the following characteristics:
- **Active Cursor** is a Read Attribute. The values read are set by paths {a} and {b}. Therefore, if Cursor 1 is the middle cursor, the cursors on either side of the middle cursor are Cursors 2 and 3.

- **Cursor Index** is a Read Attribute, in contrast to the first Attribute Node at the upper left corner of the While Loop. Its value is set by the data paths {c} and {d} that start at the Cursor Index indicator in the upper left corner of the While Loop. The result is that the control labeled **Spread** sets the value, s, which is added to and subtracted from the middle cursor index. The result is that if Cursor 1 has the index i, Cursors 2 and 3 will automatically move to the indices i-s and i+s.
- **Cursor Y** is a Write Attribute for all three attribute nodes. Therefore, the three cursors provide the voltages $v(i-s)$, $v(i)$, and $v(i+s)$, which are used to determine dV_1 and dV_2 in the calculation of the time constant that takes place in the upper right quadrant of the While Loop in Fig. M2-22.

The While Loop enclosing this part of the VI is needed to keep the VI running so it can recalculate the time constant every time you move the center cursor (Cursor 1) around.

c) Modify your VI to match Fig. M2-22. Change the **channels** control so that you acquire only one channel, the one with the exponential waveform. Set the **spread** control to some small number, like 5. Run your VI with your RC circuit and square-wave input attached.[Don't forget to turn on the **Measure tau** toggle switch.] Adjust your VI controls until you obtain a front panel screen that resembles Fig. M2-23.

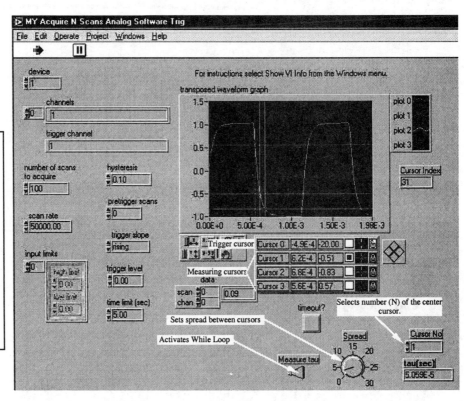

Fig. M2-23. Front Panel of the VI for measuring exponential time constant.

d) Adjust the spread control and move the center cursor until you get the three cursors to bracket the region where the exponential is falling most rapidly (similar to Fig. M2-23). Print out a copy of this screen and place it here.

Fig. M2-24.
Waveform Graph of the VI for measuring exponential time constant. Cursors covering the rising exponential.

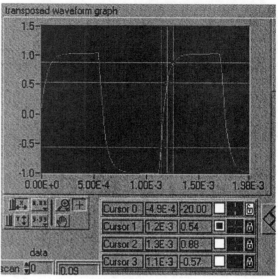

e) Using the same **spread** setting as in part d), move the center cursor until you get the three cursors to bracket the region where the exponential is **rising** most rapidly (similar to Fig. M2-24). Compare the resulting values of time constant obtained.

Fig. M2-25.
Front panel showing the case where the three cursors do not cover a region where the exponential is clearly rising or falling.

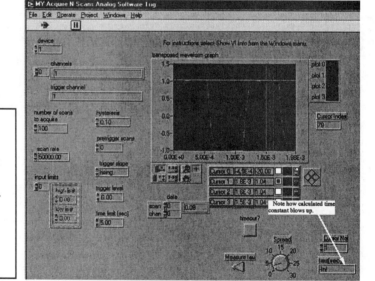

f) Drag the center cursor right and left a few points and observe how resulting value of τ changes. Record 10 different measurements of τ in the table on the following page and calculate the mean and standard deviation. **When taking your measurements, avoid situations such as that shown in Fig. M2-25, in which the signal is essentially flat.** Also, avoid letting the region between the cursors cover portions of both a falling and a rising exponential.

Record your ten time-constant measurements in this table:

Measurement No.	Time-constant (sec)
1	
2	
3	
4	
5	
6	
7	
8	
9	
10	

Calculated Mean = _____ (sec)

Calculated standard deviation = _____ (sec)

Lab Skill Exercise M2-3: Automating the averaging process.

The previous exercise shows what usually happens when you make measurements on noisy signals. The results are distributed about some average value, and this average is a better estimate of the quantity measured than any single measurement alone. No signals are completely free of noise. A detailed account of the statistics of noise in electronic systems is beyond the scope of this workbook. One way of averaging out the fluctuations due to noise in this measurement of exponential time constants is to use the **Exponential Fit .vi** available in *LabVIEW*. In this exercise, you will build a VI that performs an exponential fit on a section of data you select with two cursors.

a) Read about the the **Exponential Fit.vi** and the **Derivative x(t).vi** in the **OnLine Help**, the *LabVIEW* manual, or the *LabVIEW* textbook you are using.

Note that the **Exponential Fit .vi** cannot perform a fit to a function of the form of your signal, an exponential added to a constant term, which has the form
$$v(t) = A\exp\{-\alpha t\} + B,$$
where A and B are constants. It can perform a fit for A and α, but not for B. However, you can get around that problem by differentiating $v(t)$ with respect to time, yielding a pure exponential, with the same decay constant as before, i.e.,
$$\frac{dv}{dt} = -\alpha A \exp\{-\alpha t\}.$$
However, using the **Derivative x(t). vi** to perform the differentiation introduces a new problem: As explained in the **OnLine Help** for this VI, the differentiation is performed on an array, and the value of the derivative obtained is inaccurate at the both ends of the array.

b) Study the modification to **the Acquire N Scans Software Trig.vi** shown in Fig. M2-26 until you understand its operation.

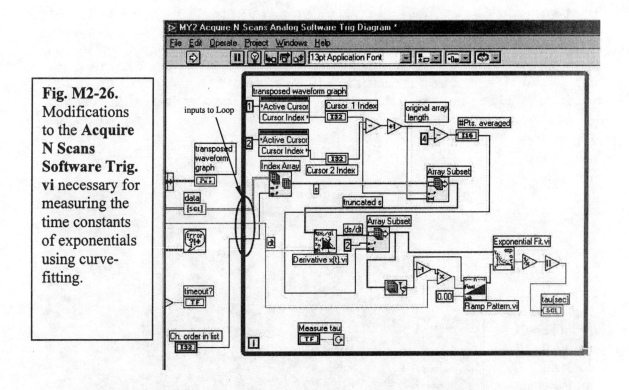

Fig. M2-26. Modifications to the **Acquire N Scans Software Trig. vi** necessary for measuring the time constants of exponentials using curve-fitting.

Inside a While Loop, which keeps the VI operating while you move the cursors around, the two cursors set the length of a one-dimensional array containing the desired signal. This waveform then:
- is extracted from the 2-dimensional array **data** using the **Ch. order in list** control, to select the right channel. For example, if the list of channels in the **channels** control on the front panel reads 2,1,0 and the signal you want to measure is on Channel 0, you would set the the **Ch. order in list** control to 2, since *LabVIEW* counts lists using the sequence 0,1,2,..... In this case if the signal you wanted were on Channel 2, you would set the **Ch. order in list** control to 0. The result of this step is a one-dimensional array, which is **number of scans to acquire** long.
- is differentiated by the **Derivative x(t) .vi.**
- has the first two and the last two samples in the differentiated signal removed, to avoid the problems the **Derivative x(t) .vi** has with inaccuracies at the ends of arrays.
- is finally fit to an exponential using the **Exponential Fit .vi**. The **Exponential Fit. vi** requires an array of times at its x input, which are associated with the array of voltages at its y input. This array of times is provided by the **Ramp Pattern.vi**.

c) Open a copy of the **Acquire N Scans Software Trig.vi,** and immediately save it as **MY2 Acquire N Scans Software Trig.**

vi. Add to your VI the modifications in Fig. M2-26. If you're still having trouble understanding how this VI works, click the **OnLine Help** for each of the VI's in the While Loop and read their descriptions.

d) Use the same RC circuit you built for the earlier exercise in this section and run your VI. Show the square-wave input to your circuit on Channel 1 and the exponential output on Channel 0. Lock Cursors 1 and 2 to Channel 1. Since you want measurements on Channel 1, set the **Channel order in list** control to 0. Set your **scan rate and number of scans to acquire** such that you get about 1-2 periods of the input and output on the graph. You should get a screen similar to Fig. M2-27.

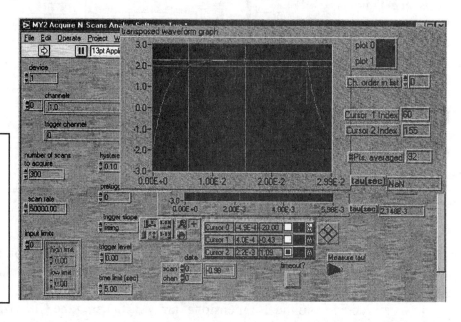

Fig. M2-27. Front Panel of the VI for time-constant measurement using curve-fitting.

Fig. M2-28. Same front panel as Fig. M2-27: Cursors now bracket a region in which the exponential to be measured is essentially flat.

Keeping **#pts averaged** constant, drag the pair of cursors left a few points and observe how the measured tau changes. Record 10 different values and calculate the mean and standard deviation. In your measurements, avoid situations such as that in Fig. M2-28, in which the section of the data where the fit is performed becomes nearly flat. Also, avoid letting the region between the cursors cover portions of both a falling and a rising exponential.

Record your ten time-constant measurements in this table:

Measurement No.	Time-constant (sec)
1	
2	
3	
4	
5	
6	
7	
8	
9	
10	

Calculated Mean = _____ (sec)

Calculated standard deviation = _____ (sec)

d) Compare your standard deviation for the 10 measurements using this method with the table of 10 measurements using the approach in the previous exercise. Which approach is more likely to yield accurate results with a single measurement? Answer here:

Chapter M2 *LabVIEW* Reference Table:

Topic	Source		
	LabVIEW Example VI's (On-line)	*Learning with LabVIEW*	*LabVIEW for Everyone*
Basics	All Fundamentals categories	Chs 1 & 2	Chs. 2-5
Structures	Fundamentals-> Structures, Graphs	Ch. 5	Ch. 6
Arrays and Graphs	Fundamentals-> Arrays, Graphs	Chs. 6 & 7	Chs. 7 & 8
Waveform Generation	DAQ Analysis->Signal Generation	Ch. 8	Chs. 10 & 11
Use of Sub-VI's		Ch. 4	Chs. 3 & 4
Waveform Acquisition	DAQ	Ch. 8	Ch. 11
Curve-fitting	Analysis->Curve Fitting	Ch. 10	

Chapter M3. Measuring Waveform Spectral Content

The earlier chapters of this workbook have shown you that the DAQ board in your computer can only make measurements directly in the time domain. However, analysis of electronic circuits is often much more convenient to perform in the frequency domain. In order to compute and display frequency domain measurements, your computer must rely on the basic transformation processes that allow you to convert data from one domain to another. These are summarized briefly below.

M3.1 Summary of Fourier Series and Transform

Many of the waveforms of interest to engineers are periodic in nature. There are some characteristics of ideal periodicity that real laboratory waveforms do not have, but it is still possible to treat them as approximately periodic. For a function of time, $v(t)$, to be truly periodic, with a period T, it must obey the relation,

$$v(t) = v(t \pm NT), \qquad (M3\text{-}1)$$

where N is any integer. This requirement is impossible for any real waveform to satisfy exactly, since it would require that the waveform be "on" for all time. However, if the real waveform satisfies Eq. (M3-1) for a fairly wide range of N, say several hundred consecutive integers, the waveform can be treated as approximately periodic.

Such an approximately periodic function can be represented as an infinite series of sine waves, having the form of a Fourier Series, which looks like

$$v(t) = A_0 + \sum_{m=1}^{\infty} \left(A_m \sin\left\{ \frac{2\pi m t}{T} + \phi_m \right\} \right). \qquad (M3\text{-}2)$$

The term, A_0, is independent of time, and can be thought of as a the dc offset of the waveform. Its value is determined by the integral,

$$A_0 = \frac{1}{T} \int_0^T v(t)\,dt, \qquad (M3\text{-}3)$$

where the interval, 0 to T, corresponds to the time interval over which the waveform has been measured.

The coefficients, A_m in Eq.(M3-2), each represent the amplitude of the Fourier component of the waveform with the frequency m/T. Since the m-th term in the summation in Eq. (M3-2) has a frequency which is m times that of the lowest non-zero frequency present, this term is called the m-th harmonic in the waveform. The term for $m=1$ is called either the *first* or the *fundamental* harmonic. The coefficient, A_m, is found from the measured waveform by the relation,

$$A_m = \left\{ \left(\frac{1}{T} \int_0^T v(t)\sin(2\pi m t/T)\,dt \right)^2 + \left(\frac{1}{T} \int_0^T v(t)\cos(2\pi m t/T)\,dt \right)^2 \right\}^{1/2}. \qquad (4)$$

In a voltage waveform, it has the dimensions of volts. The phase ϕ_m, of the m-th harmonic is given by

$$\phi_m = \arctan\left\{\frac{\int_0^T v(t)\sin(2\pi mt/T)dt}{\int_0^T v(t)\cos(2\pi mt/T)dt}\right\}. \tag{M3-5}$$

Therefore the Fourier spectrum of a periodic waveform takes the form of two lists of numbers with the index m, one list for the amplitude and another for the phase.

If the voltage waveform under consideration is not periodic, and no waveform really is, the analogous amplitude and phase spectra are not discrete series, but rather continuous functions of frequency. They represent the amplitude and the phase of a complex function,

$$F(\omega) = \int_{-\infty}^{\infty} f(t)\exp\{-j\omega t\}dt. \tag{M3-6}$$

The job of the *Virtual Bench* Digital Spectrum Analyzer (DSA), or of any other spectrum analyzer VI you might build in *LabVIEW*, is to compute and display the amplitude and or phase spectrum of any waveform captured by your DAQ board. If the VI computes the spectra of more than one waveform, it may also do comparison operations such as computing an impulse response or transfer function. Since the computer calculates these spectra from voltage waveform samples takes at a sampling rate f_s, the frequency range covered by the amplitude and phase spectra will be

$$0 \le f \le \frac{f_s}{2}. \tag{M3-7}$$

The amplitude and phase spectra are represented in the computer and on the display as plots consisting of N/2 points, where N is the total number of samples taken when capturing the waveform. This number is often called the **frame size** of the signal acquisition. Therefore, each point plotted on a graphic display by a digital spectrum analyzer represents an average over a frequency band that is f_s/N Hertz wide.

The amplitude spectrum can be visualized in a number of different ways. In the above discussion, the amplitude spectrum is proportional to the magnitude of the function, *F*, in Eq. (6), while the phase spectrum is its angle in the complex plane. On some occasions you may be interested in the frequency regions in which the signal you are measuring carries most of its power. You can generate a **power spectrum** from the amplitude spectrum just by squaring it. You can also normalize this power spectrum so that is shows **power density** per unit bandwidth (the dimensions are watt/Hz). Other units include dB scales for amplitude or for power. Since dB scales must be referenced to something, the usual reference level is 1 volt, 1 volt rms, or 1 watt, depending on the units in which the amplitude spectrum is being measured.

M3.2 Spectral measurement with the *Virtual Bench* Digital Spectrum Analyzer

Figure M3-1 shows the front panel of the Virtual Bench DSA in a typical measurement situation. The area to the right of the two graphs contains the display

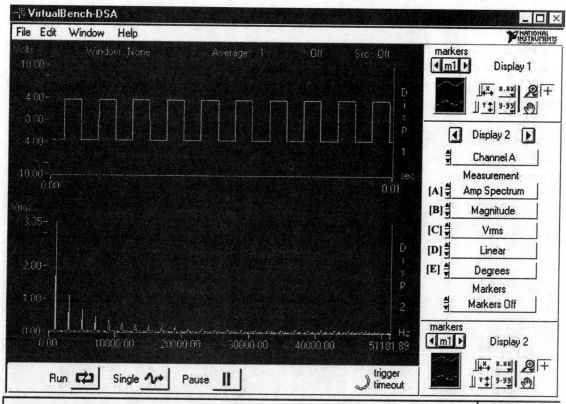

Fig. M3-1.
The front panel of the *Virtual Bench* Digital Spectrum Analyzer (DSA).

controls. The two arrows at the top of this area toggle the display controls between Display 1 (the upper graph) and Display 2 (the lower graph). After selecting a display, you can set it up using the following menu choices. The control labeled [A] selects the quantity to be displayed from among these options:

1. **Time waveform** displays the captured signal as a function of time, the same as an oscilloscope would.
2. **Amp spectrum** displays the amplitude spectrum of the captured signal as a function of frequency.
3. **Pwr spectrum** displays the power spectrum of the captured signal as a function of frequency.
4. The other options on this menu allow you to make comparisons between two different captured signals, computing a **cross-power spectrum**, **frequency response**, **coherence** function or **impulse response**. In order to use these options properly, you must take account of the fact that the two signals to be compared are sampled alternately, not simultaneously. For the basic frequency spectra measurements in this chapter, you will not need these options anyway.

The control labeled [B] toggles between **Magnitude** and **Phase**. Of course, this control has meaning only for displays where the values are complex. For example, if you selected **Time Waveform** in the control [A], above, the choice of between **Amplitude** and **Phase** has no effect.

The control labeled [C] lets you chose the units for your display. Not every choice is available for every display. The available units are volts (rms or peak amplitude), volts^2 (equivalent to power) rms or peak, the square root of spectral power density : peak or rms v/(Hz)$^{1/2}$, or spectral power density: peak or rms v^2/Hz.

The control labeled [D] lets you select a vertical scale for amplitude measurements: **Linear, dB,** or **dBm**. The control labeled [E] does the same for phase measurements: **degrees** or **radians**.

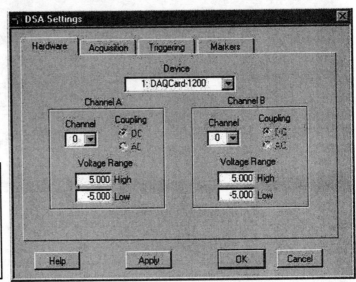

Fig. M3-2. The **Hardware** section of the **Settings** menu for the *Virtual Bench* DSA.

Before using the *Virtual Bench* DSA, you need to set up the measurement by choosing **Settings** from the **Edit** menu. Figure M3-2 shows the **Hardware** section of the DSA settings. Your DAQ device and its settings should be properly identified here, along with the input voltage range. If your lab instrumentation has already been configured, you will only need to associate up to two DAQ input channels with the DSA channels, A and B. Figure M3-3 shows the **Acquisition** section of the DSA settings. You need to make two settings here:
1. **Frame Size** is the number of samples of the waveform(s) that will be taken. To increase computation speed, the *Virtual Bench* DSA allows only an integer power of 2 to be selected.
2. **Sampling rate** is the rate at which the DAQ board takes samples from all channels in use. In the *Virtual Bench* DSA, you have either one or two channels of input. The effective sampling rate per channel is either:
    - The sampling rate, if DSA channels A and B are set to the same DAQ board input channel, OR
    - Half the sampling rate, if DSA channels A and B are set to different DAQ board input channels.

Your choices for these items will determine the horizontal axes in your displays. If you display a time waveform, the length of time displayed is

$$T_r = (framesize)/(effective sample rate).$$ (M3-8)

If you display any type of spectrum, the frequency range of the spectrum will be from zero to half the effective sampling rate per channel. Therefore, if your measurement

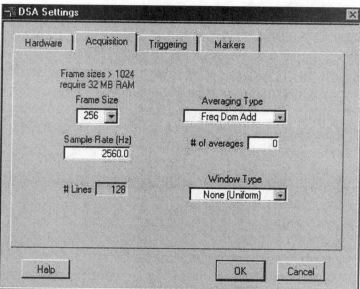

Fig. M3-3. The **Acquisition** section of the **Settings** menu for the *Virtual Bench* DSA.

involves N channels and you want to cover a frequency range from zero up to B Hz, you need to use a sampling frequency of at least

$$f_s \geq 2NB. \qquad (M3-9)$$

To illustrate the use and some of the properties of the *Virtual Bench* DSA, consider some example waveforms with well-known spectral properties:

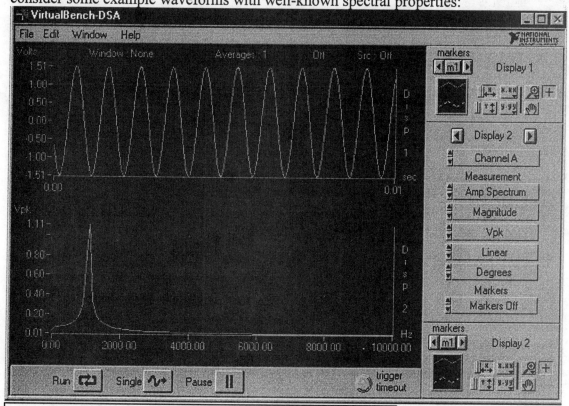

Fig. M3-4. *Virtual Bench* DSA front panel for Example 1.

Example M3-1. Sine wave (1.5 volt amplitude): Both displays in Fig. M3-4 show properties of the waveform acquired on Channel A. The upper graph, Display 1, shows the time waveform. The frame size and the sampling rate have been chosen rather arbitrarily. Note particularly that the time waveform display shows a fractional number of periods of the waveform. Display 2 shows the amplitude spectrum of this section of a sine wave. Since a sine wave is by definition a wave at a single frequency, you might well expect the amplitude spectrum to look like a single spike at the frequency of the sine wave. Although you can clearly see this spike in the amplitude spectrum, it also appears that there are some other frequencies present in the acquired waveform. These frequencies arise in the amplitude spectrum because the *Virtual Bench* DSA calculates it assuming that the waveform is periodic with a period of T_r, the overall time range of the measurement. As this periodic function repeats itself, there will be some discontinuities in the waveform every T_r seconds, and these will result in the appearance of other frequencies in the amplitude spectrum. A sine wave input to the *Virtual Bench* DSA gives an amplitude spectrum that looks nearly ideal (a single spike at the sine wave frequency) only if the sampling rate and frame size produce a T_r that is close to an integral number of periods of the measured waveform. Figure M3-5 illustrates this situation and shows why it is important to "frame" an integral number of oscillations of a periodic function within the **Frame Size** if you want to compare its spectra with those predicted by a Fourier series.

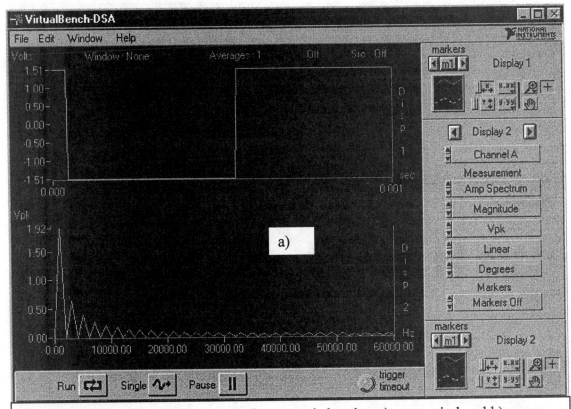

Fig. M3-5. The *Virtual Bench* DSA front panel showing a) one period and b) many periods of a square wave as the acquired waveform on Display 1, and the resulting amplitude spectra on Display 2.

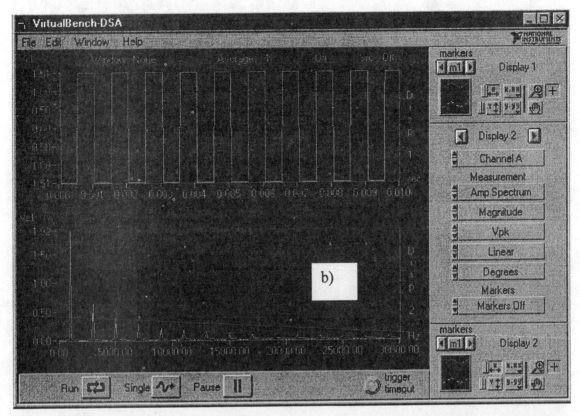

Example M3-2. One period versus many periods within the Frame Size. Figure M3-5 shows the time waveforms and the amplitude spectra for **Frame Size**s that include only one period (Fig M3-5a) and many periods (Fig. M3-5-b) of the same square wave. A Fourier series calculation on a square wave shows that it has only odd harmonics (m=1,3,5,7....) and that their amplitudes are proportional to 1/m. The previous example shows that the amplitude spectrum of a sine wave is just a spike at the sine wave frequency with an amplitude equal to that of the sine wave. Thus, a summation of sine waves such as a Fourier series should produce a set of spikes. For the square wave with a fundamental frequency of f_0, you should expect a spike one unit high at f_0, another 1/3 unit high at $3 f_0$, another 1/5 unit high at $5 f_0$, etc. The spikes corresponding to these frequencies have the correct positions and amplitudes in both Figs. M3-5a,b). The spikes in Fig. M3-5b) are narrower. This is because of the nature of the plots produced when the amplitude spectrum is measured: If N samples of the original waveform are taken at a sampling frequency f_s, then the amplitude spectrum is plotted by joining N/2 data points distributed over the frequency range, $0 \leq f \leq f_s/2$.

Figure M3-5b) results from more samples than Fig. M3-5a), so the frequency interval between harmonics is covered by more data points and the spikes look narrower.

Example M3-3. Phase spectra at all frequencies. Figure M3-6 shows the *Virtual Bench* DSA displaying the power and phase spectra of the waveform from Example M3-2. The amplitude spectrum on the upper display makes it obvious that the only significant power in this square waveform is at the frequencies f_0, $3 f_0$, $5 f_0$, etc. The relatively low power density at other frequencies results from the inevitable noise present

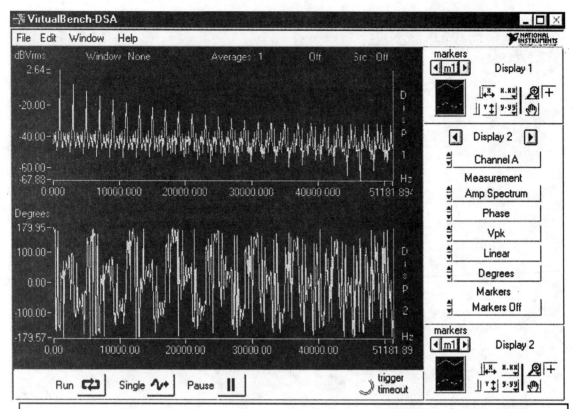

Fig M3-6. Power and phase spectra for a square wave on the *Virtual Bench* DSA.

in the circuitry. the *Virtual Bench* DSA tries to measure and display the phase of all the frequencies within its measuring range of $0 \leq f \leq f_s/2$, even if the associated amplitudes at those frequencies are negligibly small. If you took repeated measurements of the phase spectrum for a signal like this one, the phases measured for the frequencies other than the odd harmonics of f_0 would be more or less randomly distributed over all possible values. Only the phases measured at the odd harmonics of f_0 would take on their correct values (for the square wave of this case, zero) for each measurement.

```
Lab Skill Exercise M3-1: Use the Virtual Bench DSA to
measure the amplitude spectrum of a square wave.  Set up
the DSA to measure frequencies up to at least 5 times the
square-wave frequency. Select a sampling frequency and
frame size such that the time waveform display shows as
near as possible to exactly eight periods of the waveform.
Select Edit->Settings->Markers from the menu and set up
frequency markers using five harmonic markers for each
display.  Use the Markers to measure the amplitudes and
phases at 1, 3, and 5 x f₀. Display the amplitude spectrum
of the peak voltage with a linear scale on Display 1.
Display the phase spectrum in degrees on Display 2.  Line
up the markers on both displays so they are at 1, 2, 3, 4
and 5 x f₀.   Read the results from the marker locators on
```

the display and enter the results in the first two blank columns of the table below.

Harmonic	Frequency	Amplitude	Phase	Power	Power/Hz
1					
2					
3					
4					
5					

Attach your front panel print-out here:

Next, change the vertical scale to **Pwr Spec** (power spectrum measurement) and record the power carried in each of the harmonics in the **Power** column of the table above. Finally, change the vertical scale to peak voltage squared per Hz and measure the power density at harmonics 1 through 5, and record your results in the **Power/Hz** column of the table above. Since power is proportional to the square of voltage, are the ratios between harmonics in the **Amplitude**, **Power**, and **Power/Hz** columns what you expect them to be?

> Enter your answer and comments here:

Lab Skill Exercise M3-2. Study the spectral characteristics of the noise output of the **FuncGen** VI in Section M1.3.
• If you have not already done so, build the VI. If *LabVIEW* is not installed on your computer, your laboratory may have a free-standing noise generator you can use as a substitute.
• Connect the output of the noise generator to the input of the *Virtual Bench* DSA. Set the **Sampling Rate** of the DSA to about three times the **Update Rate** of the noise generator.
• Set the **trigger** to **none**.
• Display the **Time Waveform** on the upper graph and the Amplitude Spectrum on the lower graph of the DSA. Observe several runs in succession, either by pushing the **Single** button several times or by pushing the **Run** button on the DSA for a continuous display. Visually estimate the average amplitude spectrum (averaged over the runs you observed, and sketch your estimated average here:

Theoretically, the amplitude spectrum of white noise (both peak and rms) is supposed to be flat with respect to frequency. However, all real noise generators have limited bandwidth and output negligible amplitude above some limiting frequency. The upper limit to the flat region of the amplitude spectrum of the output of the noise generator is the update rate. You will only see this bandwidth limitation in your measurement results if the **Sampling Rate** for the DSA is at least *twice* the upper bandwidth limit. Theoretically, the amplitude spectrum should be flat up to this bandwidth limit and then drop to zero at higher frequencies.

Finite-time samples of real noise waveforms have spectra that are flat only if they are averaged over many runs or very long sampling times. The *Virtual Bench* DSA allows you to automate some of this averaging:

```
With your experiment wired the same way as for the
earlier parts of this exercise, choose Edit->Settings-
>Acquisition from the Virtual Bench DSA menu, and select
FreqDomAdd from the averaging choices. This option
accumulates multiple measured spectra and averages the
results of the multiple runs at each frequency. Then set
the # Averages to 20, and observe a few of the resulting
amplitude spectrum plots. Increase the # Averages
gradually until the average amplitude spectrum begins to
flatten out and approach the appearance you would predict.
(Don't expect the averaging to be a miracle cure here - -
the # Averages required to really flatten out the amplitude
spectrum would probably take longer to acquire than you
would want to wait.)
```

You might be curious about what would happen in the above exercise if you had chosen to do the averaging in the time domain instead of the frequency domain. This choice is **TimeDomAdd** in the **Settings->Acquisition** menu. Under this option, the averaging would be carried out in the time domain on each run, and the spectra of amplitude and/or phase versus frequency would be calculated only after the averaging was completed. Since white noise theoretically averages to zero (provided, of course, that you average over enough runs) the time domain average would tend toward zero and the amplitude and/or phase spectra would be meaningless.

The *Virtual Bench* DSA has many other available options and functions which increase its versatility. These features are useful primarily in more advanced electronics laboratory operations, and will not be discussed here.

M3.3 Frequency-domain measurements with *LabVIEW*

If you have *LabVIEW* installed, you can build your own Digital Spectrum Analyzer (DSA) VI and have it perform all of the functions of the *Virtual Bench* DSA. If you have done the Lab Skill Exercises in the *LabVIEW* section of Chapter M2, you have already done most of the programming required. The VI you will use in this section is an extension of the analog software trigger.vi in Section M2-2.

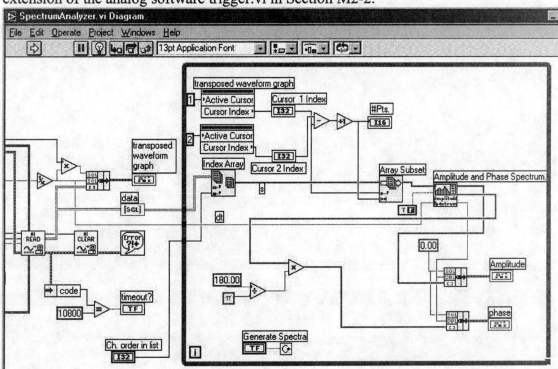

Fig. M3-7. Diagram of a *LabVIEW* VI to select a portion of an acquired waveform and measure its amplitude and phase spectra.

Figure M3-7 shows the modifications to the **Acquire N Scans Analog software Trigger.vi** from the Examples library necessary to measure the amplitude and phase spectra of a section of an acquired signal selected by the operator. The operator determines the section of waveform to be Fourier transformed using the two cursors. In Fig. M3-8a), the operator positioned the cursors on the waveform so as to include only one cycle of the waveform. In Fig. M3-8b), the cursors are positioned to include many cycles. You can see frame size effects similar to those visible in Fig. M3-5 with the *Virtual Bench* Digital Spectrum Analyzer.

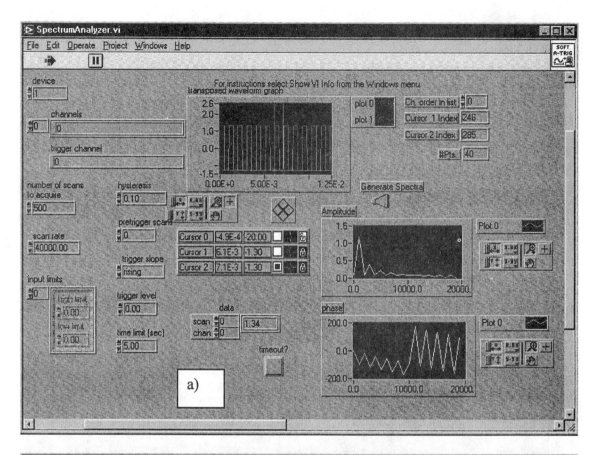

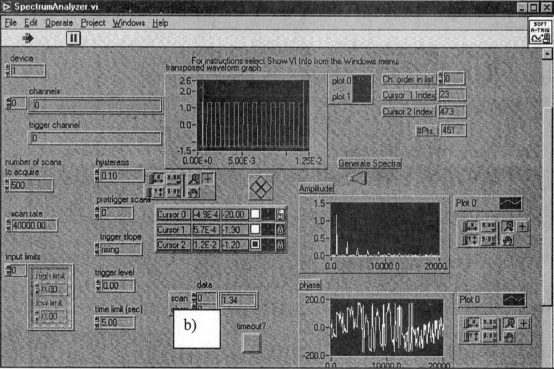

Fig. M3-8. The front panel of the **Spectrum Analyzer.vi** for a) cursors selecting a single period and b) cursors including many periods of a square wave.

Lab Skill Exercise M3-3: Using the Spectrum Analyzer VI.
• Build the **Spectrum Analyzer** VI, above.
• Test your VI using a square wave input from the *Virtual Bench* Function Generator or some other available square-wave source. Obtain an amplitude spectrum that shows at least the first five harmonics of this square wave. Attach it here:

Topic	Chapter M3 LabVIEW Reference Table: Source		
	LabVIEW Example VI's (On-line)	*Learning with* **LabVIEW**	*LabVIEW for Everyone*
Basics	All Fundamentals categories	Chs 1 & 2	Chs. 2-5
Structures	Fundamentals-> Structures, Graphs	Ch. 5	Ch. 6
Arrays and Graphs	Fundamentals-> Arrays, Graphs	Chs. 6 & 7	Chs. 7 & 8
Waveform Generation	DAQ Analysis->Signal Generation	Ch. 8	Chs. 10 & 11
Use of Sub-VI's		Ch. 4	Chs. 3 & 4
Waveform Acquisition	DAQ	Ch. 8	Ch. 11
Spectrum Analysis	Fundamentals-> Analysis->Signal and Spectrum Analyzers	Ch. 10	

Chapter M4: Measuring complex transfer function

Introduction

The concept of a transfer function is restricted to linear circuits, or at least to those whose behavior can be approximated by a linear model over a restricted range of operation. For such circuits, a sinusoidal input at a single frequency results in a sinusoidal output at the same frequency. Since the input and output are both sine waves at a single frequency, they can only be compared with each other on two bases:

- Relative amplitude. To measure an amplitude ratio, you need only observe a display of the two sine waves and calculate a ratio of the amplitudes or of the peak-to-peak voltages.
- Phase shift. To measure a phase shift, ϕ, you must determine the time interval, dT, between *equivalent points* on the two waves and compare it to their common period, T, using the relation,

$$\phi = \frac{dT}{T} \times \begin{cases} 360 (\text{degrees}) \\ 2\pi (\text{radians}) \end{cases}. \qquad (M4\text{-}1)$$

Equivalent points are usually one of the following:
- Positive-going zero-crossings,
- Negative-going zero-crossings,
- Positive peaks,
- Negative peaks.

Often, the zero-crossings (or crossings of a particular voltage level other than zero) can be located more accurately than the positive or negative peaks of a sine wave, since sine waves have zero slope at these peaks.

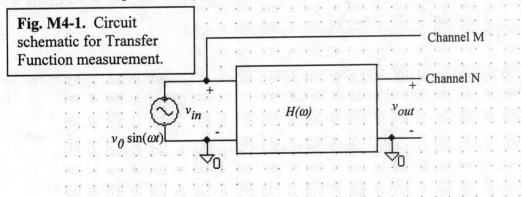

Fig. M4-1. Circuit schematic for Transfer Function measurement.

The circuit for measuring a transfer function is shown in Fig. M4-1. The "black box" whose transfer function, $H(\omega)$, is to be measured is driven by a sine wave at he frequency of interest. The input waveform is acquired on Channel M and the output on Channel N of the DAQ board. You can use either the *Virtual Bench* Oscilloscope or one of the *LabView* **Acquire N Scans ... vi.**s to accomplish this, by following the procedures in Sections M2.1 and M2.2.

M4.1 Magnitude and phase comparison with sine-wave input

Since measuring phase shift requires that you know the time relationship between two sine waves, you have to acquire these two waveforms simultaneously. Since most DAQ boards have several analog input channels, you can easily capture two waveforms.

Non-simultaneous sampling. However, a problem can arise in accurately determining a time delay between specific parts of two different waveforms. The cause is the way in which typical DAQ boards collect samples. Because a good sample-and-hold circuit is relatively expensive and space-consuming, most DAQ boards sample multiple channels in turn, the way a dealer distributes a hand of cards. In the *Virtual Bench* Oscilloscope and in the *LabView* **Example vi**'s we use in this workbook, the

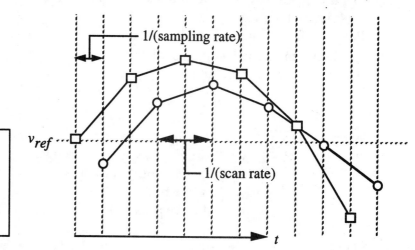

Fig. M4-2. Voltage vs. time for two sampled waveforms from a timed acquisition.

scan rate and the sample rate are related as shown in Fig. M4-2. If N is the number of channels to be scanned, then in a *timed acquisition*, the sampling rate, f_s, and the scan rate, f_{sc}, are related by the equation,

$$f_s = Nf_{sc}. \tag{M4-2}$$

When you use the *Virtual Bench* Oscilloscope, you control the sampling rate and the number of channels, which sets the scan rate. Conversely, when you use the *LabView* **Acquire N scan...vi**'s, you control the scan rate and the number of channels. In both cases, the DAQ board's capabilities set an upper limit on the sampling rate: *LabView* and *Virtual Bench* will both warn you if you attempt to set this rate too high for your DAQ board.

All that the DAQ instrumentation "knows" about either of the sampled voltages in Fig. M4-2 is the set of number pairs (v, t) represented by the squares for the first channel and the circles for the second. In order to draw the waveforms as correctly as possible, the *Virtual Bench* Oscilloscope horizontally displaces each successive channel in the scan by $(1/f_s)$ with respect to its predecessor in the scan order. *Virtual Bench* Oscilloscope does this automatically when drawing waveforms on the screen. *LabView* does not, and this requires some corrective programming measures to be discussed later in this Chapter.

Phase measurement with Cursors. If you have used the cursors on either the *Virtual Bench* Oscilloscope or a *LabView* **Acquire N scan... vi,** you already know that you can't set a cursor between two sample points on the display. The cursor will only move from one sample to another. Thus, if you want to use a cursor to locate the time of the first positive-going zero crossing of the waveform whose samples are the circles in Fig. M4-2, you will have to be content with an error of the order $(1/f_s)$ in that measurement.

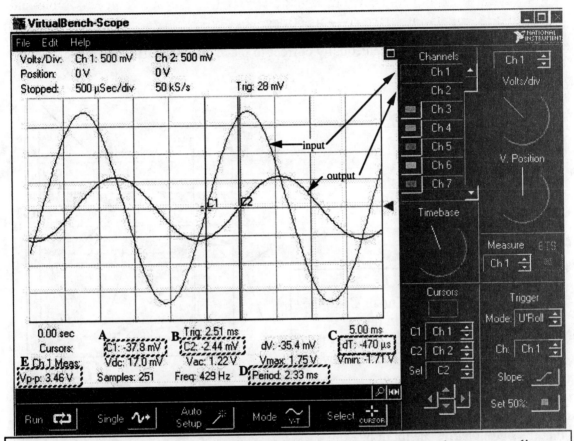

Fig. M4-3. The *Virtual Bench* Oscilloscope displaying the input and output to a linear circuit whose transfer function is to be measured.

Figure M4-3 shows a *Virtual Bench* Oscilloscope screen that displays a sinusoidal input (Channel 1) and the resulting output (Channel 2). You can measure the phase shift between the input and output with the following procedure:
- Display the input waveform on Channel 1 and the output on Channel 2.
- Set the Measure control to provide measurements on Channel 1. Be sure that the peak-to peak amplitude, Vp-p , is included in the measurements that are displayed.
- Use the Cursor control to attach Cursor C1 to Channel 1 and Cursor C2 to Channel 2
- Position C1 and C2 as near as possible to the zero-crossings of the input and output. You want the **magnitudes** (ignore the signs) of the voltages in Boxes A and B as small as possible.
- Record the **magnitude** (ignore the sign) of the time delay, dT, from Box C of Fig. M4-3. This is also *dT* in the phase shift calculation of Eq. (1).

- Record the waveform Period from Box D of Fig. M4-3. This is T in the phase shift calculation.
- You now have all the information necessary to calculate the phase shift using Eq. (M4-1).

The question of the sign of the phase shift, i.e., whether it is a lead or a lag, needs some discussion here. The way the cursors are used in Fig. M2-4 to calculate dT, it seems obvious that the output *lags* the input by

$$\phi = \frac{dT}{T} \times 360 \, \text{deg}. \tag{M4-3}$$

However, you would have been equally correct to place C1 on a different zero-crossing

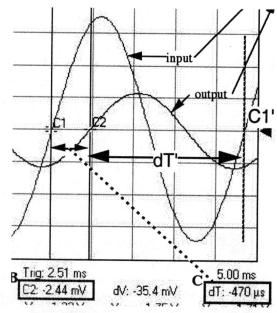

Fig. M4-4.
The effect of moving Cursor C1 to the position C1', resulting in the conclusion that the zero crossing of the output leads that of the input by a time dT'.

of the input waveform, as shown in Fig. M4-4. This would have resulted in a conclusion that the output *leads* the input by

$$\phi' = \frac{dT'}{T} \times 360 \, \text{deg}. \tag{M4-4}$$

If you perform measurements of both ϕ and ϕ' for a real pair of sine waves, you will always find that they sum to 360 deg or 2π radians. Since the transfer function is the ratio of complex amplitudes of output to input, the phase angle, α, in

$$H(\omega) = |H(\omega)| \exp\{j\alpha(\omega)\} \tag{M4-5}$$

is either:
1. $-\phi$, if you have measured the output as *lagging* the input as shown in Fig. M4-3.
2. $+\phi'$, if you have measured the output as *leading* the input as shown in Fig. M4-4.

Both of these choices result in the same value for $H(\omega)$ as a complex number.

Phase measurement error. It's worthwhile to estimate the phase measurement error in this procedure so that you can decide if you want to undertake an even more complex measurement.

The mean square error in this phase measurement can be written as

$$\varepsilon_\phi^2 = \left(\frac{\partial \phi}{\partial T}\right)^2 \varepsilon_T^2 + \left(\frac{\partial \phi}{\partial (dT)}\right)^2 \varepsilon_{dT}^2, \tag{M4-5}$$

where ε_T and ε_{dT} are respectively the root-mean square errors in the period, T, and the time-delay, dT. The equation for calculating the phase shift, ϕ, gives the required partial derivatives as

$$\begin{aligned} \frac{\partial \phi}{\partial T} &= -360 \frac{dT}{T^2} \\ \frac{\partial \phi}{\partial (dT)} &= \frac{360}{T} \end{aligned} \tag{M4-6}$$

It remains to estimate numerical values for ε_T and ε_{dT}, the errors in the quantities directly measured. Recall that you read the period, T, directly off the *Virtual Bench* Oscilloscope as 2.33 ms. You can think of this value as $2.330 \pm .005$ ms, since the number for the display was rounded off to three significant figures. Therefore, the root mean square error in measuring the period is

$$\varepsilon_T = 0.005 ms = 5 \times 10^{-6} s.$$

To estimate the root mean square error in measuring the time delay, dT, note from Fig. M4-3 that the number of samples per second for each trace on the display is given as 50kS/s (see immediately above the graph, in the center), which means the time between successive samples is 20μs. Therefore, a reasonable estimate for the root mean square error in measuring the time delay is

$$\varepsilon_{dT} = 10 \mu s = 1 \times 10^{-5} s.$$

The total error in Eq. (M4-5) is the sum of two terms, the first for the error in measuring the period, and the second for the error in measuring the delay. Before calculating the total error, it is interesting to look at the contribution made by these two terms. Plugging in the numbers for the period error only gives a rms error of 0.156 deq, while calculating the rms delay error only gives an error of 1.55 deg, about 10 x larger. These rms errors sum to give a total measurement error of 1.553 deg for the phase shift, which is not bad at all.

`Prep Exercise M4-1.` Verify the above rms error calculations for this phase shift measurement.

`Prep Exercise M4-2.` Suppose you had the same oscilloscope screen data as in Fig. M4-2, but your DAQ board were sampling much more slowly, and number of samples per second

for each trace were only 5kS/s. Estimate the new rms error
in measuring the phase shift. ε = _____deg.

Magnitude comparison. To complete the transfer function measurement at this frequency, you need a comparison of the peak-to-peak amplitudes of the two sine waves.

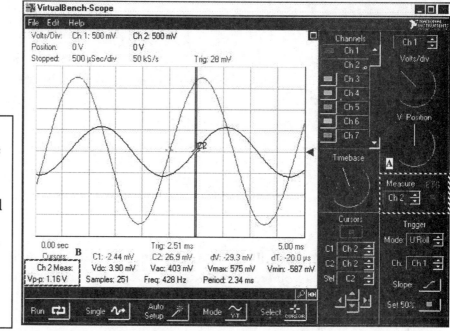

Fig. M4-5. The same scope trace as Fig. M4-3, with the Measure control set on Channel 2 to provide the peak-to-peak voltage of the output waveform.

You can do this by switching the Measure control (see Box A in Fig. M4-5) to Channel 2. Then the measured data for Channel 2 appears below the graph as shown in Fig. M4-5. The magnitude of the transfer function is just the ratio of the Vp-p measurements (Box B in Fig. M4-5 and Box E in Fig. M4-3).

To measure a transfer function vs. frequency, just repeat the above steps for each frequency for which you want a data point. You will need to take data for enough frequencies to get a realistic picture of the desired curves of $H(\omega)$ and $\alpha(\omega)$. The problem of getting enough data points for an accurate plot is analogous to that of sampling signals in the time domain: for these frequency-domain measurements, you need more closely spaced data points in frequency regions where the magnitude and/or phase are changing rapidly. Repeating the above measurement process for enough different frequencies can be time-consuming. Consequently, if you have *LabView* installed, it's worthwhile to consider writing some Virtual Instruments which automate part or all of this process, as discussed below.

M4.2. Magnitude and phase measurement in *LabVIEW*

This section describes two ways to measure a transfer function with *LabVIEW* VI's, avoiding some of the labor associated with the process of collecting a series of data points directly from time-domain measurements described above. The first method automates the calculations in the amplitude and phase comparisons at each frequency selected for measurement. The second method supplies to the circuit an input that contains a broad range of frequencies and uses Fourier Transforms to calculate the transfer function versus frequency from a single time-domain measurement.

M4.2.1 Point-by-point vector voltmeter (FDomain.vi or SFTF.vi)

A vector voltmeter compares the magnitude and phases of two sine-wave voltages at the same frequency. One of these voltages is the input to a two-terminal-pair network like Fig. M4-1 and the other voltage is the output. The transfer function is just the ratio of the two complex amplitudes. If the instrument uses Fourier analysis on the input and the output, then the input does not need to be an ideal sine wave. However, for the VI described below to work well, one frequency in the input should be significantly stronger than all the rest. You can use the *Virtual Bench* Function Generator, or a free-standing function generator as the input source for transfer function measurements with this virtual vector voltmeter.

Start by opening the **Acquire N Scans – Multistart.vi** in the Examples->daq-> anlogin-> anlogin.llb directory.

```
•Examine this VI's diagram, review the VI Info, and
complete Prep Exercise M2-6, if you have not already done
so.
```

This VI can acquire and display the input and output time waveforms for any circuit connected like Fig. M4-1. You want to enhance it so you can determine the peak (largest-amplitude) frequency in these waveforms, and then calculate a ratio of the complex amplitudes of the output and input at that frequency. If you built a sub-VI in *LabVIEW* to accomplish that purpose, the front panel might look like Fig. M4-6.

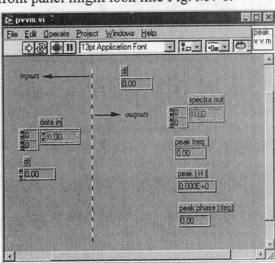

Fig. M4-6.
Front Panel of a sub-VI to calculate transfer function at the peak frequency of a given input and output.

The inputs to this VI consist of the sampling interval, *dt.*, and a two-dimensional array containing the sampled input and output waveforms. The outputs are:
- a two-dimensional array containing the amplitude spectra of input and output (spectra out),
- the frequency interval, *df*, for this array (analogous to *dt* for a time-domain measurement),
- the peak frequency (the component of the output frequency spectrum with the largest amplitude),
- the ratio of output to input amplitude at this peak frequency (**peak |H|**). This is the magnitude of the transfer function to be measured;
- the phase difference between the output and the input at the peak frequency in degrees (**peak phase (deg)**).

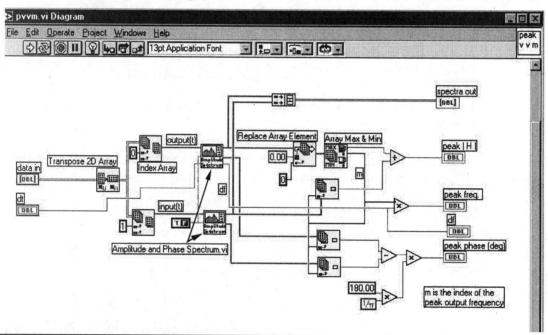

Fig. M4-7.
A Sub-VI for measuring transfer function magnitude and phase at the strongest frequency present in the output waveform.

Figure M4-7 is a sub-VI diagram showing how the outputs are generated from the inputs in *LabVIEW*. We assume the incoming time waveform data to be in the form that the *LabVIEW* analog input VI's usually present it: each row represents one scan across the input channels, and each column represents the successive scan values for a particular channel. Since the other *LabVIEW* VI's expect time waveforms to be presented as rows instead of columns, the first thing the VI must do is transpose the input data array, as shown at the left of Fig. M4-7. Next the two **Index Array .vi**'s separate the input and the output waveforms, providing them as inputs to two **Amplitude and Phase Spectrum. vi**'s. These VI's have two array outputs: the amplitude (upper) and the phase (lower) spectra of the input time waveform. These spectra are functions of frequency over the range,

$$0 \le f \le \frac{1}{2dt},$$

that you would expect from the Whittaker-Shannon sampling theorem. Then, the **Replace Array Element.vi** sets the dc component of the output spectrum to zero. Sometimes, you may want to measure small-signal transfer function parameters (such as the small-signal gain of a transistor amplifier) where the signal of interest has a large dc offset. This zeroing of the dc component insures that the **Array Max & Min.vi** will never select zero as the peak component in the output spectrum. Instead, it will pick out the highest amplitude from the remaining frequency components and identify its index, m, in the array. The frequency corresponding to that index is mdf, and its value is output as the **peak frequency** by the sub-VI. Finally, the amplitudes at that frequency are divided and the phases are subtracted in order to provide the **peak |H|** and the **peak phase** outputs from the sub-VI.

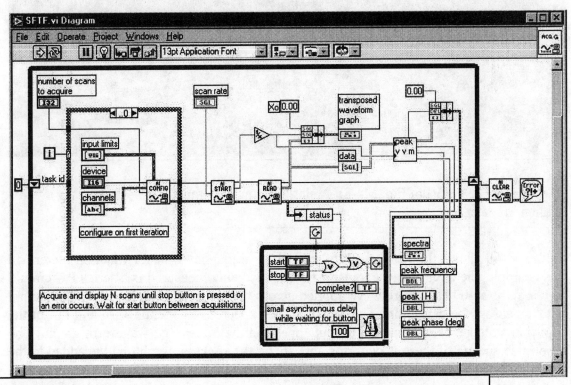

Fig. M4-8.
The **peak v v m** sub-VI wired into the **Acquire N Scans – Multistart .vi** to make a single-frequency transfer function meter.

Figure M4-8 shows how to wire this sub-VI into the **Acquire N Scans – Multistart .vi** to make a single-frequency transfer function meter. The outputs of the sub-VI (shown in the diagram as a **peak vvm.vi**) are brought out to the front panel as digital indicators for **peak frequency**, **peak |H|**, and **peak phase**. An added waveform graph displays the amplitude spectra of the input and output as functions of frequency. To obtain transfer function data at different frequencies, you simply change the peak frequency of the input waveform you are supplying and tabulate the results. This VI does the magnitude and phase calculations for you.

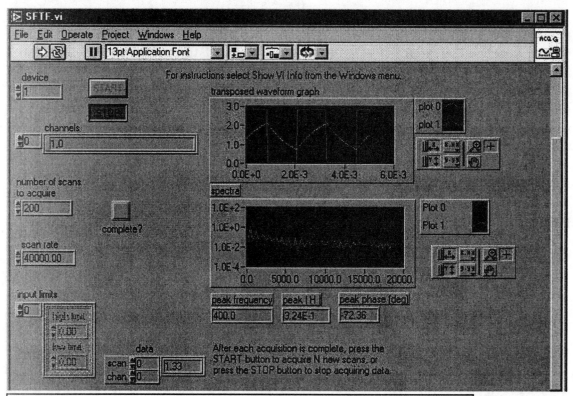

Fig. M4-9.
Front panel of the VI of Fig. M4-8: A single-frequency transfer function meter.

Figure M4-9 shows the front panel of this single frequency transfer function meter. In this example, the circuit whose transfer function is being measured is a series RC circuit. The input is a 400 Hz square wave applied across the series RC combination. The output is taken across the capacitor. The output is the expected exponential rise and fall of the capacitor voltage. The input and output appear on the transposed waveform graph. The amplitude spectra of these waveforms appears on the **spectra** graph that we added to the **Acquire N Scans – Multistart .vi.** Close inspection of the amplitude spectra would reveal that both have maxima at 400 Hz. The digital indicators we added for **peak frequency, peak |H|** and **peak phase** show the measured value of the transfer function at 400 Hz.

With this VI, your input voltage does not have to be a clean sine wave as it would if you measured the transfer function using the time-domain methods of Section M4.1. Figure M4-10 shows the time waveforms, amplitude spectra, and resulting measured transfer function parameters for the same RC circuit using sinusoidal and clipped triangle inputs. Note that the transfer function results are very similar.

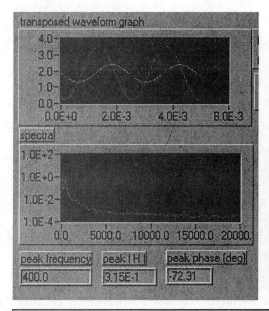 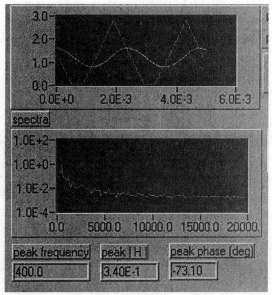

Fig. M4-10.
Outputs from the single-frequency transfer function meter measuring a series RC circuit using sinusoidal (left) and clipped triangle (right) waveforms as inputs.

Lab Skill Exercise M4-1. Open *LabVIEW* and build the sub-VI whose front panel and diagram are shown in Figs. M4-6 and M4-7 respectively. Wire the connector for this sub-VI so you can use it in another VI.

• Next, construct your own single-frequency transfer function meter by wiring your sub-VI into the **Acquire N Scans Multistart.vi**, as shown in Fig. M4-8. Arrange your front panel like Fig. M4-9.

• Select some standard values for a resistor and capacitor for your circuit to test. Enter your selections in this table, along with your calculated values for the magnitude and phase of the transfer function.

Resistor value (Ω)	
Capacitor value (farads)	
Measurement frequency (Hz)	
Calculated $|H|$	
Calculated phase shift, $\angle H$ (deg)	

• Use your VI to measure $|H|$ and $\angle H$ with different excitation waveforms at the **same measurement frequency**. Enter your results on the following table:

Excitation type:	|H|	∠H (deg)
Sine wave		
Triangle wave		
Square wave		

- Print out a front-panel of your VI that shows the results of your transfer function measurement using a square wave excitation, and attach it here:

M4.2.2. Instant data: Broad-band transfer function measurement

The preceding section showed how you can measure transfer functions at a single frequency by looking at a single frequency value in the output of the **Transfer Function.vi**. This section now asks and answers the question: why not excite the circuit whose transfer function is to be measured with a waveform that contains *all* the frequencies within a range of interest, and generate curves for the magnitude and phase as function of frequency with one measurement? Let's approach this measurement as a top-down design problem in *LabVIEW*. If you need more background on Fourier Transforms and how they are used to obtain frequency spectra of signals, the following references may be useful:

Topic	Nilsson & Riedell	Sedra & Smith
Fourier series & transform	Chs. 17 & 18	Chs. 7 & 8
Transfer Function	Section 14.4	

The VI you build to measure a broad-band transfer function will need to do the following:

1. Generate a broad-band analog signal to be used as input to the circuit. White noise, by definition, carries equal amplitudes at all frequencies. However, because of the random nature of noise, a noise signal lasting for a finite length of time will only approximate this.
2. Acquire the time waveform at the output to the circuit, *simultaneously* with generating the broad-band input. Simultaneous analog output and input from a DAQ board requires some special programming steps in *LabVIEW*, but there are some example VI's that do this in the **Examples-> daq -> anlog_io.llb** library. Depending on the DAQ board you have in your computer, you may have to start with a different VI from this library, but the library does have an example VI for most boards. Once you pick out the correct example VI as a starting point, the steps that follow are similar in all cases.
3. Compute the transfer function and display the results as graphs of transfer function magnitude, $|H|$, and phase, α, versus frequency. The **Transfer Function.vi** in the analysis library makes this computation. As usual, the Whitaker-Shannon sampling theorem governs the frequency range for which you can obtain measurements: from zero up to half the sampling frequency (*scan frequency* in the case of multi-channel signal acquisition).

The above are necessary specifications for broad-band transfer function measurements. One more feature is highly desirable: frequency-domain averaging over a number of repeated measurements. Element m (remember, the first element of an array in *LabVIEW* is element 0, not element 1) of the arrays output by the **Transfer Function.vi** represents the measured magnitude (or phase) of the transfer function at a frequency mdf, where df is the frequency increment between data points. Averaging many measurements at each frequency will reduce errors due to noise. The VI you build will include calculation of a running average of transfer function measurements over each run since the VI was started, until you tell the VI to stop.

Figure M4-11 shows how the front panel of this broad-band transfer function measurement VI might look. The upper waveform graph labeled **wg1** shows the noise

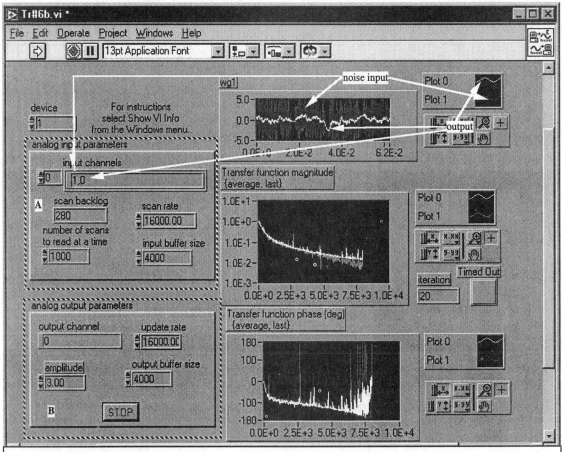

Fig. M4-11. Front panel of a VI for broad-band transfer function measurement.

input and the resulting output time waveforms as acquired by the DAQ board. The lower two waveform graphs show the transfer function magnitude or phase versus frequency, with the last measurements in dark gray, and the running averages in white. The number if iterations used to calculate the averaged data is shown to the right of the graph for transfer function magnitude.

The VI whose front panel is shown above was built by modifying the **Simultaneous AI/AO Buffered (E-Series MIO).vi** in the **anlog_io.llb** library.

`Prep Exercise M4-3.` Open the `Simultaneous AI/AO Buffered (E-Series MIO).vi,` pull down the `Show VI info` menu and read about this VI.

Figure M4-12 shows the diagram of the **Simultaneous AI/AO Buffered (E-Series MIO).vi** as it appears when you open it from the *LabVIEW* examples library.

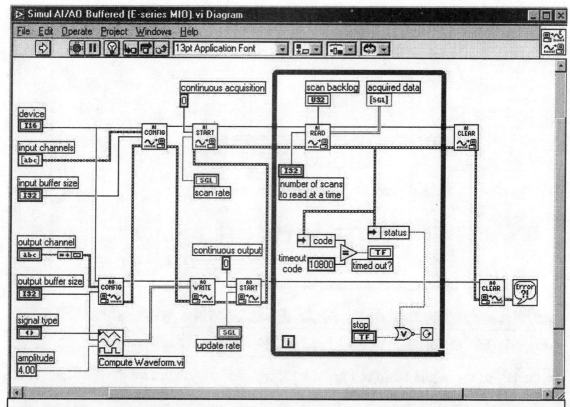

Fig. M4.12. A simultaneous AI/AO VI from **the anlog_io.llb.** Several are available: the correct one to use depends on which DAQ board or card is in your computer.

When you are finished modifying this VI, it will look like Fig. M4-13, on the following page. The modifications are summarized as follows:

- The While Loop has been expanded to the left, to allow repeated simultaneous analog output and input.
- The output waveform generator has been changed to provide a Gaussian noise input to the circuit with adjustable amplitude.
- A sub-VI has been added to calculate and plot instantaneous and averaged magnitudes and phases for the transfer function measurements.

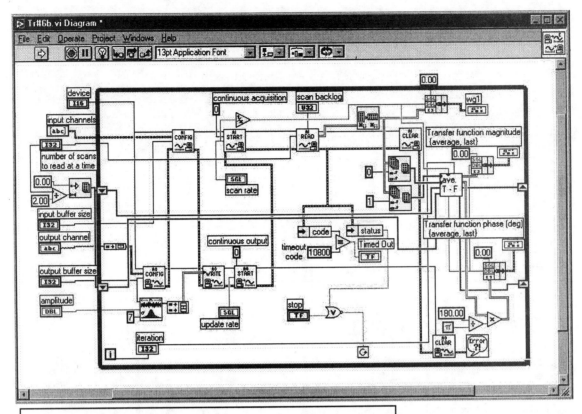

Fig. M4-13. The diagram of the VI in Fig. M4-11.

To the left of the While Loop in Fig. M4-13, two shift registers are initialized with zero values. These shift registers contain one-dimensional arrays *half* the size of **the number of scans to read at a time**. If the VI reads N scans each time and plots two N-point time waveforms on the waveform graph labeled **wg1** in Fig. M4-11, then the amplitude and phase spectra calculated from these waveforms are arrays of size N/2.

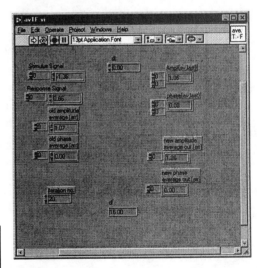

Fig. M4-14.
Front Panel of the **ave TF** sub-VI.

Figure M4-14 shows the front panel of the **ave TF** sub-VI at the right of Fig. M4-13. It averages a freshly calculated transfer function (amplitude and phase) with all the previously acquired transfer functions from earlier iterations. Although this sub-VI handles arrays, you can best understand how it works by considering the averaging of a sequence of numbers from a measurement: Suppose four measurements have already been taken (during iterations 0-3 of a while loop) and each of those measurements has a value of 4.0. The average of these four measurements is obviously 4.0. Now suppose the next measurement (the fifth one, taken on iteration 4 of the while loop) is 9.0. The new average is 5.0 (the average of four 4.0's followed by a single 9.0).

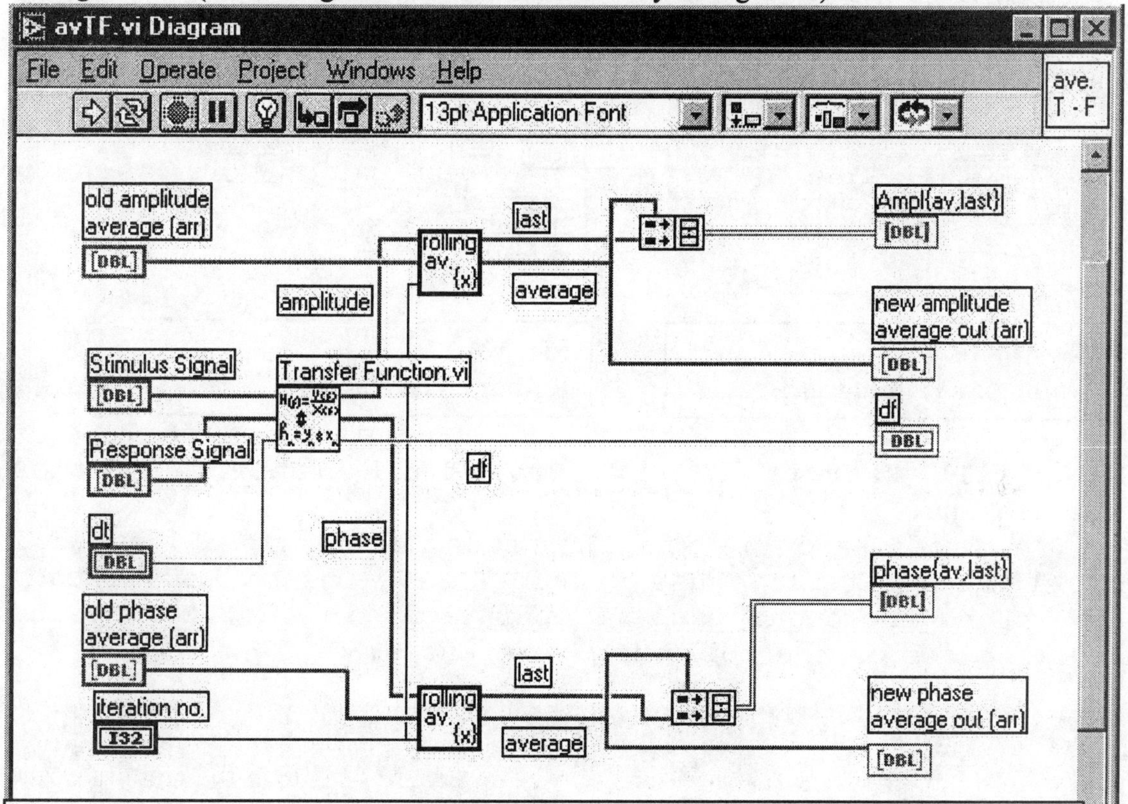

Fig. M4-15. Diagram of the aveTF sub-VI.

Figure M4-15 shows the diagram of the **ave TF** sub-VI. The newly acquired **Stimulus Signal** and **Response Signal**, along with the sampling interval **dt** are input to the **Transfer Function.vi** from the Analysis library of *LabVIEW*. Its output arrays for amplitude and phase are fed to two **rolling-av{x}** sub-VI's, who average this new result with all the previous ones from earlier iterations of the While Loop of the main VI. Each **rolling-av{x}.vi** then passes a two dimensional array out to the main VI. The first row of each of these arrays contains the latest averaged amplitude or phase, updated with the last data acquisition. The second row of each of these arrays contains only the most recently acquired set of amplitude or phase data.

There is one more sub-VI *within* the **ave TF** sub-VI that remains to be examined: the **rolling-av{x}.vi.** Figure M4-16 shows its front panel, and Fig. M4-17 its diagram. Its operation is self-explanatory.

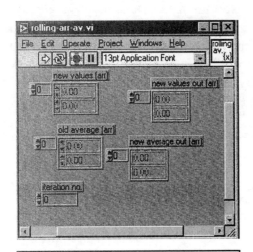

**Fig M4-16.
Rolling-av.vi** front panel.

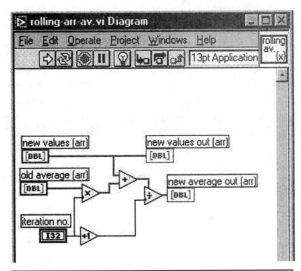

**Fig. M4-17.
Rolling-av.vi** diagram

`Lab Skill Exercise M4-2.` Broad-band transfer function measurement.
• The Broad-band transfer function measuring VI you will build makes extensive use of sub-VI's. Unless you already understand thoroughly sub-VI operations in *LabVIEW*, review at least one of the following references:

Topic	*LabVIEW* User Manual	Learning with *LabVIEW* (Bishop)	*LabVIEW* for Everyone (Wells & Travis)
Sub-VI creation and wiring	Ch. 2	Ch. 4	Ch. 5 p. 128ff.

• Build the Broad-band Transfer Function VI described in this section.
1. Begin by building the necessary sub-VI's:
    • **rolling-av{x}.vi**, and
    • **ave TF.vi.**
2. Wire their connectors so that they can be used as sub-VI's, with all of their inputs and outputs accessible at their terminals.
3. Open the **Simultaneous AO/AI . . .vi** in the **anlog_io.llb** that applies to your DAQ board. Save a copy immediately under another **xxx.vi** name. Modify it to match Fig. M4-13.
4. that that its diagram looks exactly like Fig. M4-13.

- **Lab Skill Exercise M4-3**: Test your VI from the previous exercise on a series RC network. Go through the following steps:

Scan rate (Hz)	
Update rate (Hz)	
Frequency range for measurement (Hz)	
Resistor (Ω)	
Capacitor (F)	
$f_0 = (RC)^{-1}$ (Hz)	

1. Select the **scan rate** on your VI front panel. Set this rate to slightly less than half the maximum sampling rate for your DAQ board, since you are using two analog input channels into this VI. Enter your selected scan rate in the table above.
2. Select the **update rate** on your VI front panel. **For the VI to work properly your update and scan rates should be equal.** If *LabVIEW* refuses to let you set the update rate at this value, you will need to pick a lower scan rate in Step 1, and the set the update rate to the same value. Enter your selected update rate in the table above. Revise your table entry from Step 1 if necessary.
3. Calculate the frequency range (determined by the Sampling Theorem) over which you expect to get measurements from this VI. Enter that range in the table above.
4. Select standard values for R and C to produce a time-constant whose reciprocal lies near the middle of the frequency range from Step 3. Enter your resistance and capacitance values, and the resulting frequency corresponding to $f_b = (RC)^{-1}$ in the table above.
5. Wire up your VI to your test circuit. Run the VI several times, stopping at about 20 iterations of the measurement, and adjust the **number of scans to read at a time**, and the **input buffer size**, until your averaged amplitude and phase displays look as you would expect for a simple RC low-pass filter such as this circuit. Fig. M4-11 shows how the outputs should look for an RC low-pass filter.
6. After you have adjusted the VI as in Step 5 so that your curves look fairly smooth, change the horizontal axes of the amplitude and phase graphs to a log scale and run the VI again. Observe how the averaged data slowly settles to a reasonably smooth curve. Print out a copy of the VI

front panel showing this data and attach it in the space provided.

Attach printout here:

You may wonder why the phase measurements toward the high end of your frequency range seem to stay quite "noisy" even after the averaging has been allowed to run for many iterations. The answer to this question lies in the amplitude graph on the front panel. Note that this higher frequency region is the same frequency range in which the amplitude of the transfer function becomes small. Thus the output voltage being measured by the DAQ board is so small that its measurement is no longer very precise. The **TransferFunction.vi** in your broadband measurement VI tries to calculate a phase associated with each frequency, even though it the signal it's receiving carries negligible power at that frequency. This behavior is similar to that of the *Virtual Bench* DSA in Fig. M3-6, which calculates values for the phase spectrum at each frequency within its measuring range even if the signal is very weak. The phase of a frequency component that has essentially zero amplitude is meaningless.

This VI has one other interesting and potentially confusing characteristic. If you increase the **update rate** to a value significantly larger than the **scan rate**, your data becomes much more "noisy", and may not come close to giving the results you expect. This happens because the voltage waveforms at the input and the output are not sampled simultaneously, and the one that is sampled later is the result of the circuit being driven with a different input. However, if the scan rate and update rate are equal, both input and output voltages get sampled during one update period for the input, which makes the comparison between them valid.

Optional Lab Exercise: Try increasing the update rate to about twice the scan rate, and watch this measurement fail.

Topic	Chapter M4 *LabVIEW* Reference Table: Source		
	LabVIEW Example VI's (On-line)	*Learning with LabVIEW*	*LabVIEW for Everyone*
Basics	All Fundamentals categories	Chs 1 & 2	Chs. 2-5
Structures	Fundamentals-> Structures, Graphs	Ch. 5	Ch. 6
Arrays and Graphs	Fundamentals-> Arrays, Graphs	Chs. 6 & 7	Chs. 7 & 8
Waveform Generation	DAQ Analysis->Signal Generation	Ch. 8	Chs. 10 & 11
Use of Sub-VI's		Ch. 4	Chs. 3 & 4
Waveform Acquisition	DAQ	Ch. 8	Ch. 11
Spectrum Analysis	Fundamentals-> Analysis->Signal and Spectrum Analyzers	Ch. 10	
Simultaneous Analog Input/Output	DAQ-> Simultaneous Analog I/O		

Chapter M5. Measuring Current and Impedance

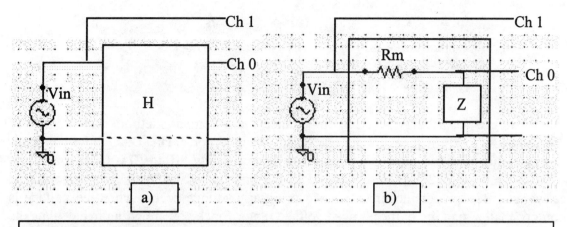

Fig. M5-1. a) transfer function measurement. b) measurement of unknown impedance, Z, with a known measuring resistor, Rm.

Introduction

Wiring a circuit for impedance measurement is no different from wiring it for transfer function measurement as described in Chapter M4. The voltage waveforms acquired are exactly the same. Only the processing of the measurements is different. Figure M5-1 illustrates this similarity. Figure M5-1a) shows a two-port network wired for transfer function measurement. Figure M5-1b) makes the assumption that the two-port network takes a specific form, namely a series combination of a measuring resistor and a circuit whose impedance is to be measured. Both measurements require acquisition of the voltages V_i and V_o, on two measurement channels connected to either:
- Two input channels of the *Virtual Bench* Oscilloscope, or
- Two analog input channels of a *LabVIEW* VI.

Impedance is given by the complex amplitude ratio,

$$Z(\omega) = \frac{V(\omega)}{I(\omega)} = \frac{|V|}{|I|}\exp\{j\alpha\}, \qquad (\text{M5-1})$$

where α is the phase difference between the voltage and the current at the frequency of interest. Fig. M5-1 shows that the waveform for the voltage across the unknown Z is $v_o(t)$ and that the waveform for the current through the unknown Z is

$$i(t) = \frac{v_i(t) - v_o(t)}{R_m} \qquad (\text{M5-2})$$

Therefore, regardless of what software the computer may use to acquire the two voltage waveforms in Fig. M5-1, it must subtract the output voltage from the input voltage and divide by a known measuring resistance in order to measure a current. This process can be automated in *Virtual Bench* as well as in *LabVIEW*.

The organization of this chapter follows that of Chapter M4 on Transfer Function Measurement. The three methods described below for impedance measurement are based on the three transfer function measurement schemes described in Chapter M4. For a review of impedance concepts and transfer function measurement methods, the following references may be useful:

Topic	Reference(s)
Impedance concept	Nilsson & Riedel Sec. 9.4
Transfer function measurements	This Workbook, Ch. M4

Before describing the three impedance measurement methods below, a brief discussion of current measurement accuracy is in order. The DAQ board in your computer quantizes the allowed input voltage range (usually –10 to 10 volts or sometimes –5 to 5 volts) into 2^M levels, where M is the bit accuracy of your analog-to-digital converter. Best percent accuracy in measuring any voltage occurs when the voltage is not too small a fraction of the allowed voltage range. Since you are going to calculate the current through the unknown impedance in Fig. M5-1 from two measured voltages, you also want the voltage drop across the measuring resistor to be not too small a fraction of the allowed voltage range. From Fig. M5-1 it is clear that the voltage drop across the measuring resistor and the voltage across the unknown impedance will be of roughly equal magnitude if R_m and $|Z|$ are roughly equal. In such a case, if the input amplitude is about 10 volts, the current through the measuring resistor produces a voltage drop of about 5 volts whose magnitude and phase can be compared with the voltage across the unknown impedance, also about 5 volts. Since impedance can vary by orders of magnitude over the frequency range of interest, you must compromise in your choice of a value for R_m. The rule of thumb is: R_m should be large enough so that the currents you expect in the unknown impedance will produce a measurable voltage drop, but not so large that nearly all the input voltage appears across R_m instead of the unknown impedance.

Prep exercise M5-1: You can supply an input voltage amplitude of 5 v to a circuit for measuring an impedance you don't know. You suspect the unknown impedance has a range of magnitudes from 10-50 KΩ over the frequency range of interest. Select a good value of R_m to use in impedance measurements. Estimate the range of amplitudes for the voltage drop across R_m and the voltage across the unknown impedance, and enter your answers in the table:

R_m (ohms)	
Range of voltage drop across R_m	
Range of voltage across Z	

M5.1. Measuring Impedance with the *Virtual Bench* Oscilloscope

For oscilloscope measurement of impedance at a single frequency, you need to:
  • drive the circuit of Fig. M5-1 with a sine wave at the frequency at which you want to measure the impedance.

- cause the oscilloscope to display two waveforms: the voltage across the unknown impedance and the current through it.

The first step requires only that you connect a suitable function generator to the input of the circuit in Fig. M5-1 and adjust its frequency and amplitude. For the second step you will need to set up the **Math Channel** of the *Virtual Bench* Oscilloscope. To do this, select **Channel Settings** from the **Edit** menu on the front panel. Click on the tab labeled **Math**. You will see a dialog box like Fig. M5-2. Type in a mathematical expression that gives the voltage difference between the channel connected to the input of

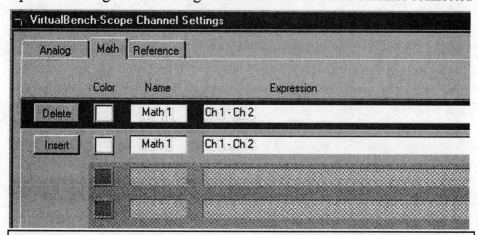

Fig. M5-2. Configuring the *Virtual Bench* Oscilloscope Math Channel

Fig. M5-1, and the channel connected to the output. The voltage waveform displayed on the scope screen when you select the Math Channel for display is now

$$v_{math}(t) = v_{in}(t) - v_{out}(t) = R_m i(t), \quad (M5-3)$$

which means that you can calculate the current amplitude by dividing the math channel voltage amplitude by R_m. Then the desired impedance is just

$$Z = R_m \frac{|V_{out}|}{|V_{math}|} \exp(j\phi), \quad (M5-4)$$

where:
- $|V_{out}|$ is the amplitude of the sine wave displayed on the *Virtual Bench* Oscilloscope channel connected to the output voltage,
- $|V_{math}|$ is the amplitude of the sine wave displayed on the *Virtual Bench* Oscilloscope Math Channel, and
- ϕ is the phase difference between the output voltage and the **Math Channel** voltage (positive if leading, negative if lagging). You measure this phase difference and compare the two amplitudes with *exactly* the same procedure as in Section M4.1.

M5.2. Impedance measurement at a single frequency using *LabVIEW*

Figure M5-1 clearly shows that the only difference between transfer function and impedance measurements lies in how the data is processed after it is acquired. If you use *LabVIEW* to write a sub-VI that will take as inputs two voltage waveforms (input and output) and the value of a measuring resistor, and produce as outputs two waveforms corresponding to the voltage, *v(t)*, across and current, *i(t)*, through the unknown impedance, you can then wire that sub-VI into any transfer function-measuring VI from Chapter M4.

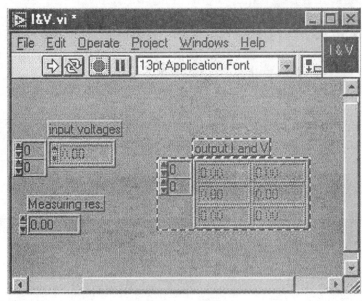

Fig. M5-3. The **I&V** sub-VI front panel

Such a sub-VI would transform an input 2-D array (the input and output voltage **waveforms**) into another 2-D array (the current and voltage waveforms), so you would expect its front panel to look like Fig. M5-3.

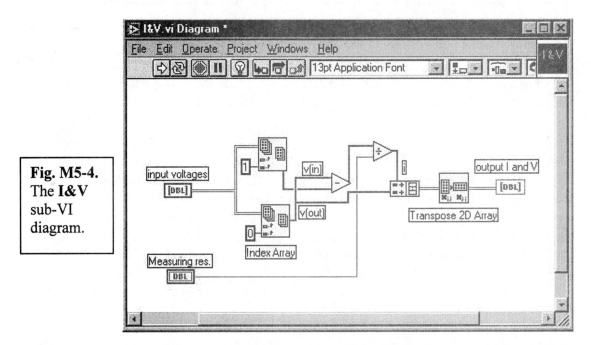

Fig. M5-4. The **I&V** sub-VI diagram.

The calculation of *i(t)* follows Eq. (M5-2). Figure M5-4 shows how you might perform this calculation in *LabVIEW*. You may or may not want to include the **Transpose Array.vi** just before the output in Fig. M5-4, to get your data ready for graphic output. If **your data array needs to be transposed before being graphed, you can also do that later, elsewhere in the VI diagram.**

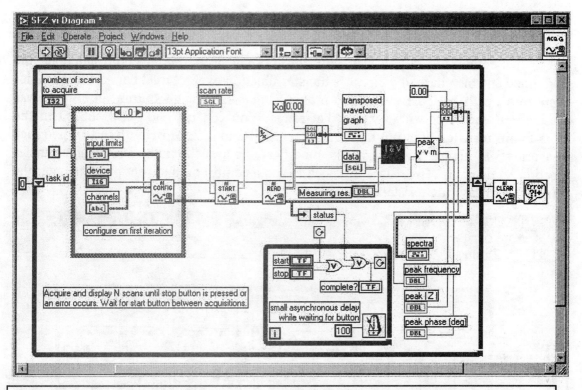

Fig. M5-5. Using the **I&V** sub-VI in the single-frequency transfer function VI of Chapter M4 to create a single-frequency impedance-measuring VI.

Calculating the impedance is just like calculating the transfer function, if you treat the current waveform as the stimulus and the voltage waveform as the response. Feed the current and voltage waveforms into the stimulus and response inputs respectively of the **TransferFunction.vi**, and its outputs will be the magnitude and phase of the impedance. Figure M5-5 shows how to place the **I&V** sub-VI into the single-frequency transfer function diagram from Section M4-2 (see Fig. M4-8) so that the magnitude and phase at the indicators are those of the impedance being measured. Note that one more control for entering the value of the measuring resistance, R_m, needs to be added to the front panel. Aside from these changes, this VI is just like the one for measuring transfer function at a single frequency. It requires that the input signal driving the circuit have one dominant frequency (whose amplitude spectrum component is larger than all the others). Therefore sine, square, and triangle-wave inputs work equally well.

M5.3. Broadband impedance measurement using *LabVIEW*

In Section M4-2, you saw how the **TransferFunction.vi** could calculate the magnitude and the phase of a circuit's transfer function over a broad band of frequencies from zero up to half the scan rate, provided that significant output amplitude was available at each frequency within that range. You can use the **I&V** sub-VI in the **Broadband transfer function** VI from Section M4-2 to make a **Broadband Impedance** measuring VI. Figure M5-6 shows the modifications to make in the Broadband transfer function VI diagram. Here also, you need to add a control on the front panel to enter the value of the measuring resistance, R_m.

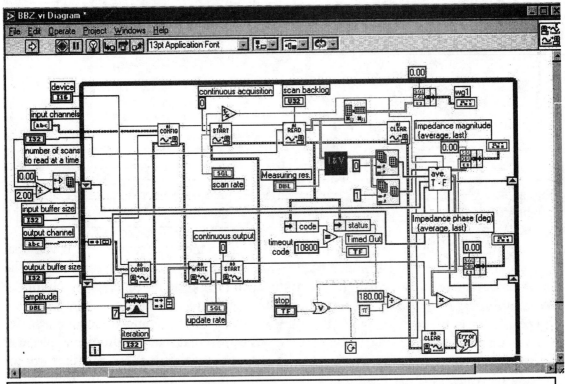

Fig. M5-6. Using the **I&V** sub-VI in the broadband transfer function VI to create a broadband impedance measuring VI.

In summary, you can easily modify any of the schemes in Chapter M4 for measuring the transfer function of a circuit so that they measure its impedance instead. You just need to incorporate an additional step which converts the input and output voltage waveforms into waveforms representing the voltage across, and the current through, the circuit whose impedance is to be measured. You should also carefully select the measuring resistor large enough that the current through it generates a measurable voltage drop, but small enough that a measurable voltage appears at the output of the measuring circuit.

M5.4 Measuring trans-impedance

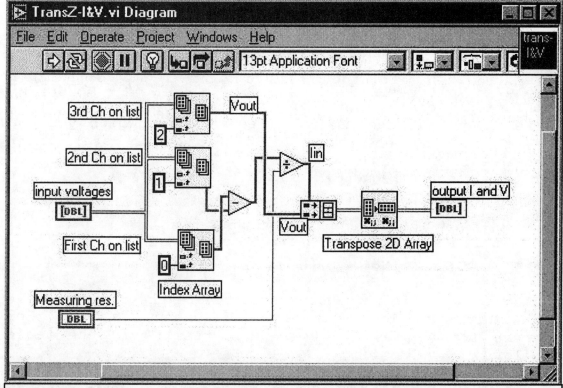

Fig. M5-7. Diagram of the TransZ I&V sub-VI. Substituting this VI for the I&V sub-VI in the broad-band impedance VI lets you measure trans-impedance.

Trans-impedance is defined for a two-port circuit as the ratio the complex amplitudes of *output* voltage to *input* current, with the output terminals open-circuited. To measure it, you still need two data channels on the input side of the circuit to measure the current through a measuring resistor. You also need a third channel of analog input to measure the output voltage. You calculate the **trans**-impedance by comparing amplitudes and phases of *output* voltage to *input* current. On the *Virtual Bench* Oscilloscope, display two waveforms:
1. the difference in voltages between the two ends of the measuring resistor, computed on the Math Channel, and
2. the output voltage.

Then compare magnitude and phase of voltage and current just as if you were measuring an ordinary impedance.

In *LabVIEW*, you can make a simple modification to the **I&V** sub-VI, and then use it in either the single-frequency or the broad-band impedance VI's described earlier in this chapter. The modification is shown in Fig. M5-7. A third input channel, which has acquired the **output** voltage, is substituted for the **input** voltage in computing the impedance. No other changes are necessary, except adding this new channel to the list of analog input channels on the main VI front panel. Just substitute the **trans I&V** sub-VI for the original **I&V** sub-VI, and you have an instrument for measuring trans-impedance.

Lab Exercise M5-1. Measure the impedance of a parallel resistor-capacitor combination over as wide a range of frequency as you can:

- Build the Single-frequency Impedance VI, the Broad-band Impedance VI and their sub-VI's.

- Calculate the magnitude of the impedance you expect in the middle of the range of frequencies you can measure. (Remember that the maximum frequency for which you can take impedance measurements is half the scan rate) Enter that magnitude in this table:

Max. measurement frequency (Hz)	
Resistor value (Ω)	
Capacitor value (F)	
$|Z|$ at half Max. measurement frequency (Ω)	
Measuring resistor value (Ω)	

- Choose a set of frequencies that covers the range from 10 Hz to the maximum measurement frequency, above. For these frequencies, fill in your results in the table below:

Freq. (Hz)	Calculated Z		Z from Single Freq. Measurement		Z from Broad-band Measurement	
	Mag.	Phase	Mag.	Phase	Mag.	Phase

M5.5. Measuring current gain

The trans-impedance measurement can also measure current gain. The trans-impedance measurement requires an open circuit at the output. To measure current gain, replace that open circuit with a known load resistor, R_L, and perform the trans-impedance measurement exactly as in the last section. Divide your measured trans-impedance by the load resistance to get the current gain.

Topic	Chapter M5 *LabVIEW* Reference Table: Source		
	LabVIEW Example VI's (On-line)	*Learning with* **LabVIEW**	*LabVIEW for Everyone*
Basics	All Fundamentals categories	Chs 1 & 2	Chs. 2-5
Structures	Fundamentals-> Structures, Graphs	Ch. 5	Ch. 6
Arrays and Graphs	Fundamentals-> Arrays, Graphs	Chs. 6 & 7	Chs. 7 & 8
Waveform Generation	DAQ Analysis->Signal Generation	Ch. 8	Chs. 10 & 11
Use of Sub-VI's		Ch. 4	Chs. 3 & 4
Waveform Acquisition	DAQ	Ch. 8	Ch. 11
Spectrum Analysis	Fundamentals-> Analysis->Signal and Spectrum Analyzers	Ch. 10	
Simultaneous Analog Input/Output	DAQ-> Simultaneous Analog I/O		

Chapter M6: Measuring Current-voltage (*i-v*) Characteristics for Non-linear devices

Introduction

In order to describe fully the behavior of a nonlinear device such as a diode, bipolar junction transistor (BJT), or field-effect transistor (FET), its manufacturer will usually provide a set of one or more characteristic curves of current versus voltage. These characteristic curves are often part of the data sheet supplied with every device. Many of these data sheets for common diodes, transistors, and other electronic devices are now available over the internet from the manufacturer's web site. In spite of this widespread availability of device data from the manufacturer, you may need to measure the current-voltage (*i-v*) characteristic curves for a particular device in the laboratory for one or both of the following reasons:

1. Certain aspects of the characteristic curves for a device may not be well-controlled in the manufacturing process. For example, in junction FET's, the value of gate voltage necessary to obtain a given saturation drain current may vary by as much as 25-50% from device to device. The data provided by the manufacturer will describe the typical, or average, device but cannot be assumed true for every device.
2. Measuring the characteristic curves for the device you are using may be the only way to know if it is functioning properly. Semiconductor junctions and FET gate oxide layers can be permanently damaged by accidental applications of voltages with too large magnitude or with the wrong polarity. This damage may leave no visible evidence; only an unwanted change in the *i-v* characteristic curves that may cause the circuit in which you are using the device not to operate as you expect.

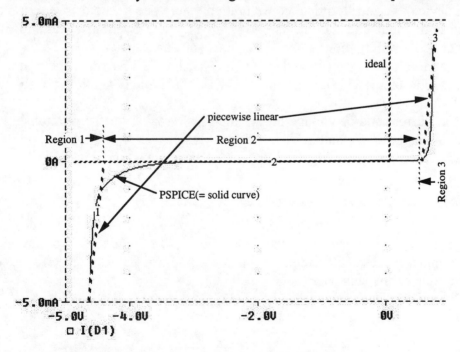

Fig. M6-1. The *i-v* characteristic curve for a D1N750 Zener diode, showing the piecewise linear regions.

For two-terminal devices such as diodes, the characteristic *i-v* curve is a single-valued function relating the current through the device (*i* on the vertical axis) to the voltage applied between the two terminals. Figure M6-1 shows an example of such a curve for a Zener diode. You may already be familiar with the use of Zener diodes in DC voltage regulators. In this application, the diode is operated in the nearly vertical part of the curve at the lower left of Fig. M6-1, where the voltage remains nearly constant over a wide range of current.

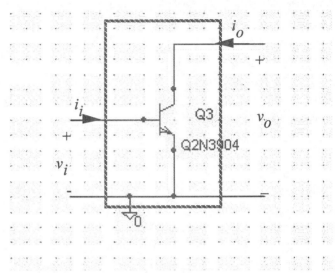

Fig. M6-2. Identification of input and output voltages and currents for a BJT in the common-emitter configuration.

For three-terminal devices such as BJT's and FET's, the *i-v* characteristics are, in general, families of curves. A device may have a set of *output* characteristic curves and another set of *input* characteristic curves. This concept results from thinking of the three-terminal device as a four-terminal configuration where one of the device terminals is *common* to both input and output, as shown in Fig. M6-2 for the example of a common-emitter BJT configuration. From Fig. M6-2, you can see that:
1. Current and voltage cannot be independent of each other, on either the output or the input side of the configuration. They have a defined relationship to each other.
2. However, current and/or voltage on the input side may alter the relationship of current to voltage on the output side, and vice versa.

These observations lead to expressing the output current characteristic curves as

$$i_o = i_o(v_o, i_i), \tag{M6-1}$$

if input current is the independent input variable. An alternative is

$$i_o = i_o(v_o, v_i), \tag{M6-2}$$

if input voltage is the independent input variable. Figure M6-2 allows identification of the input and output voltages and currents with particular terminals of the specific device as shown in the table on the next page.

General parameter	Symbol	Common Emitter BJT Parameter	Symbol
Input current	i_i	Base current	i_B
Input voltage	v_i	Base-to-emitter voltage	v_{BE}
Output current	i_o	Collector current	i_C
Output voltage	v_o	Collector-to-emitter voltage	v_{CE}

Conventionally, the output characteristics of the BJT in the common-emitter configuration are given in the form of Eq. (M6-1) and plotted as a family of curves showing collector current versus collector-to-emitter voltage with base current as a parameter. An example of output characteristics for a BJT is shown in Fig. M6-3.

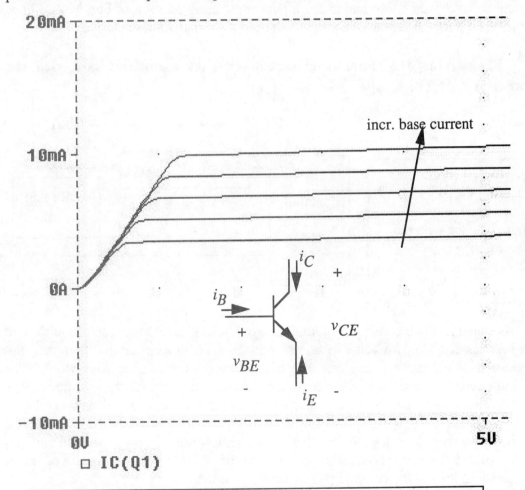

Fig. M6-3. Typical BJT output characteristic curves.

For an FET in the common source configuration, the output characteristics are conventionally given in the form of Eq. (2). In this case the identification of the input and output voltages and currents is as shown in the table below.

General parameter	Symbol	Common Source FET Parameter	Symbol
Input current	i_i	Gate current	i_G
Input voltage	v_i	Gate-to-source voltage	v_{GS}
Output current	i_o	Drain current	i_D
Output voltage	v_o	Drain-to-source voltage	v_{DS}

The above configurations are perhaps the most widely used ones for these devices, but other configurations are also useful. Clearly, any terminal of the three available ones can be chosen as the common terminal in the input-output configuration of Fig. M6-2. Each choice will result in a different set of input and output i-v characteristic curves.

M6.1. Measuring i-v characteristics of two-terminal devices with the *Virtual Bench* Oscilloscope

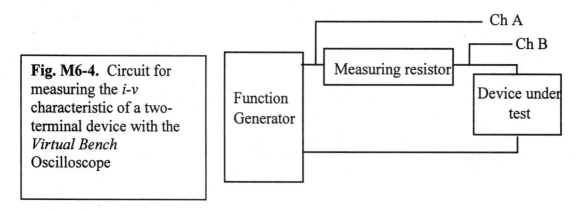

Fig. M6-4. Circuit for measuring the i-v characteristic of a two-terminal device with the *Virtual Bench* Oscilloscope

As stated in Chapter M5, measuring a current into a device or circuit requires that you add a resistor of known value in series with the device or circuit and then measure the voltage drop across this resistor. The basic idea is illustrated in Fig. M6-4. The measuring circuit is the same as the one used to measure transfer function or impedance shown in earlier chapters of this workbook. The measuring resistor, R_m, does two jobs:

1. It produces a voltage drop in a known proportionality to the current in the device being measured. Since the *Virtual Bench* Oscilloscope can only measure voltages down to a few mv, you must choose a value for R_m that will produce a voltage of al least 10 mv at the currents you expect in your device.
2. It limits the current that can flow in the device being measured. In Fig. M6-4, if the device acted as a short circuit, the maximum current would be limited to V_{in}/R_m by the measuring resistor. Many semiconductor devices can be destroyed by excessive currents. So, choosing a measuring resistor large enough to limit current below maximum values given in the manufacturer's specifications is good practice.

Both these considerations might seem to indicate that R_m should be as large as possible. However, to be useful, *i-v* characteristics like those in Figs. M6-1 or M6-3 need to extend over a reasonable voltage range on the horizontal axis. The series combination of the measuring resistor and the device being measured act as a voltage divider for the input source. If the measuring resistor is too large, it will be difficult to get much voltage across the device being measured, unless you increase the input voltage to a very large value.

Once you have selected the measuring resistor, you must set up the input voltage source so that you get the voltage you want on your *i-v* plot. For many nonlinear devices, not all values of voltage are of equal interest. For example, the Zener diode, whose *i-v* characteristic is shown in Fig. M6-1, is rarely operated at positive voltages, so a reasonable range of voltages for measurement might be $-5v<v<0.5v$. From Fig. M6-1, the maximum current to be expected in that voltage range is 5ma. For good measurement accuracy, this 5ma should produce about 1 volt when measured through the resistance, R_m. This one volt should be added to the 5.5 volt range desired for measurement, since the voltage source will need to supply the voltage drop across R_m as well as that across the device being measured. Often, you will not have enough information about the device being measured to determine before the measurement exactly what the amplitude and DC offset of the input waveform should be. Once the *i-v* characteristic is on your *Virtual Bench* Oscilloscope screen, you can adjust the amplitude and DC offset of the input voltage until the voltage range you want is covered by the measurement. A triangle or a sine wave works best as the input waveform for this measurement. A square wave spends too much time at the extremes of the voltage range and not enough time in the range between extremes. In Fig. M6-4, the Math Channel waveform is the voltage difference between the two ends of the measuring resistor. The other channel displayed is the voltage across the device being measured. To display voltage and current versus each other on the *Virtual Bench* Oscilloscope, the scope must operated in the X-Y mode, instead of the more usual V-T mode.

If you rely on the kind of model for nonlinear electronic devices that results in an *i-v* characteristic, you are forced to ignore effects like junction capacitance. The junction capacitance contributes to the current a term proportional to $C\, dv/dt$. Thus the measured *i-v* characteristic on the oscilloscope takes on a slightly different shape during the rising part of the input voltage cycle that it does on the falling part. If you should see a measured *i-v* characteristic that appears to be a closed curve instead of a line, it's probably due to parasitic capacitance. Reducing the frequency of your input waveform should eliminate this problem in your measurements.

Lab Exercise M6-1. Obtain a Zener diode of known Zener voltage, V_z, and measure its *i-v* characteristic over the voltage range, $-1.1V_z \leq v \leq 0.5V_z$. Enter your selected value for the measuring resistor here: $R_m =$ _____ ohms

Print out a copy of your Oscilloscope screen and attach it in the space provided.

M6-3. Curve-tracing for three terminal devices with *LabVIEW*

Conventional oscilloscopes can't store several *i-v* characteristic curves and display them all together. Consequently, measuring the output characteristics of a three-terminal device and obtaining a set of curves like Fig, M6-3 requires a different instrument. Conventional free-standing curve tracers measure one member of the family of curves at a time and then display them together on the screen. A *LabVIEW* VI can be

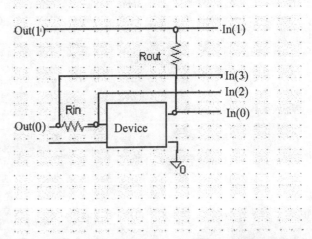

Fig. M6-5. Measuring circuit for the output characteristic curves of a three-terminal device having one terminal common to the input and the output.

written to accomplish the same thing. This VI would operate on the measurement circuit shown in Fig. M6-5. Two analog output channels [Out(0) and Out(1)] from the DAQ board provide the required voltage levels to the input and output sides of the three-terminal device configuration. Four analog input channels [In(3)- In(0)] provide enough information to calculate the input and output voltages and currents, if the two measuring resistor values, R_{in} and R_{out}, are known.

Building this curve-tracer VI provides an excellent example of how the *LabVIEW* programming environment not only promotes the good practice top-down program design, but practically forces you into it. Start with the basic requirement that this VI will have to simultaneously provide analog output for the excitation voltages, v_{inp} and v_{outp}, and measure the four channels of analog input in Fig. M6-6. This requirement suggests that you modify one of the VI's in *LabVIEW's* Examples->daq->anlog_io library. Choose the one appropriate for the DAQ board in your computer.

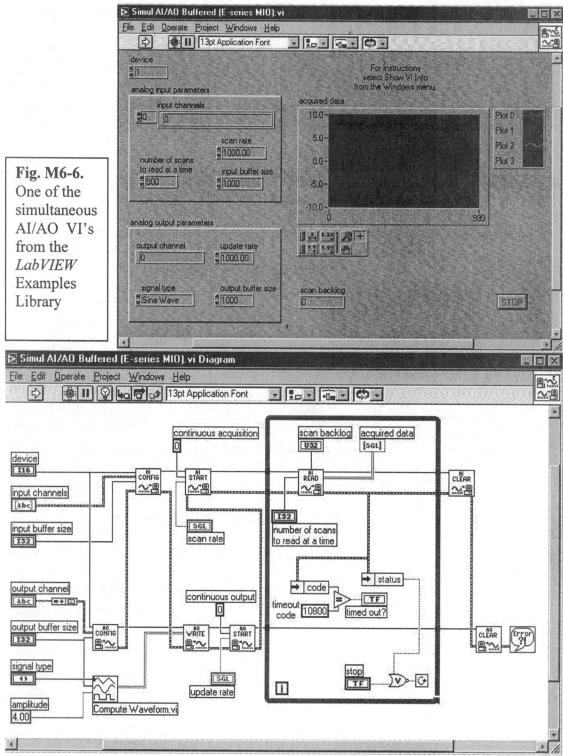

Fig. M6-6. One of the simultaneous AI/AO VI's from the *LabVIEW* Examples Library

Figure M6-6 shows the front panel and diagram for the **SimulAO/AI Buffered...
.vi** from that library.

From the nature of the characteristic curves you want to measure, the following specs for your new curve-tracer VI are self-evident:

1. Instead of displaying the acquired waveforms directly as a function of time, you will have to use these waveforms to calculate curves of current versus voltage, so your output should be an **X-Y Graph** instead of the **Waveform Graph** supplied with the example VI.
2. Instead of continuously acquiring data, you want only enough data to calculate a small number of curves. For practical reasons, four curves are sufficient for most measurements. Therefore, instead of the **While Loop** in the example that causes the analog output and input to continue indefinitely, you want a **For Loop** which executes four times and then stops and allows the i-v characteristic curves to be calculated and displayed.
3. Instead of independently setting the input and output update and scan rates and buffer sizes, you want the excitation and data acquisition synchronized with each other. Therefore the scan rate should equal the update rate, and the output and input buffer sizes should be equal. Also, when **AI Read.vi** reads the acquired data, the number of scans it reads should be equal to the buffer sizes.
4. The conventional definition of characteristic curves requires that either the input voltage or the input current should be constant over the entire range of output voltages measured. Since there is no guarantee that this will happen in a real experiment, you should monitor the input voltage and current for each curve, so that you know whether or not to believe the data on the main X-Y graph. Two small **Waveform Charts**, one for input current and one for input voltage should be sufficient for this purpose.
5. Finally, this curve tracer needs to know what values of measuring resistors, R_{in} and R_{out}, you are using in the measurement, so two controls that allow you to enter these numbers need to be added to the front panel.

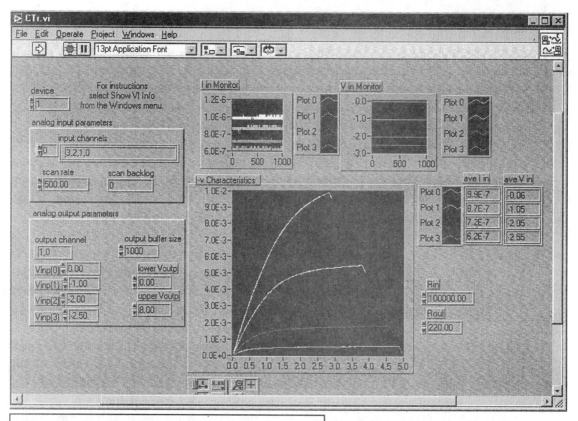

Fig. M6-7. Front panel of a curve-tracer VI.

Figure M6-7 shows how the front panel for such a VI might look. The following modifications have been made to the front panel of Fig. M6-6:
1. The **update rate** has been set equal to the **scan rate**, and only the **scan rate** control remains.
2. The **input buffer size** has been set equal to the **output buffer size**, and only the **output buffer size** control remains.
3. Digital controls, **Vinp(0)** ... **Vinp(3)**, to allow you to select the four values of v_{inp} to be applied on the input side of the device have been added.
4. Digital controls, **Rin** and **Rout**, setting the values for the measuring resistors, as well as the upper and lower ends, **upper Voutp** and **lower Voutp**, of the range for v_{outp} have been added.
5. Two small waveform graphs, **I in monitor** and **V in monitor**, show respectively the input current (into the device), i_{in}, and input voltage, v_{in}, at the input terminal of the device under test, as functions of time. For proper interpretation of the i-v characteristic curves, at least one of these two sets of curves should be constant over the whole range of the data acquisition. In Fig. M6-7, the **V in monitor** shows that v_{in} is constant for each member of the family of i-v curves. The measurement shown in Fig. M6-7 is for a junction FET (JFET). For this type of device, the output characteristics are in the form of drain current (the vertical axis in the X-Y graph labeled **I-v characteristics** in Fig. M6-7) versus drain-to-source voltage (the horizontal axis), with gate voltage as a parameter. In the range of

drain-to-source voltages shown in the graph in Fig. M6-7, and with gate voltages corresponding roughly to the values selected in the controls for **Vinp(0)** . . .**Vinp(3)**, the input current through the gate is expected to be very small. The input currents monitored by the **I in monitor** graph bear out this expectation. One reason the input currents look noisy on this monitor is that they are very small, and therefore difficult for the DAQ board to measure with great precision. In this particular example, if you had already taken the measurement shown on the front panel for this JFET and you wanted more accurrate measurements of the input current, you would increase the value of the measuring resistor, R_{in}, in your circuit, so that these small gate currents would generate larger voltage drops to measure.

6. Two array indicators, **ave I in** and **ave Vin**, to the right of the main output X-Y Graph, **I-v characteristics**, show the measured values of i_{in} and v_{in} averaged over the data for each member of the family of curves. If the monitors for input voltage and/or current indicate these quantities are constant over each curve, then the averages shown in these indicators are good values for i_{in} and v_{in}. So you can easily tell which curve color goes with each set of i_{in} and v_{in} values in these indicators, it is a good idea to line them up with the curve colors on the main graph, as done in Fig. M6-7. Also, choose the same set of colors for Plots 0 through 3 on all the graphs. Since you may use this VI on many different devices, without knowing in advance what values of voltage and currents to expect, choose the **Autoscale** option on both axes on all three graphs on this front panel.

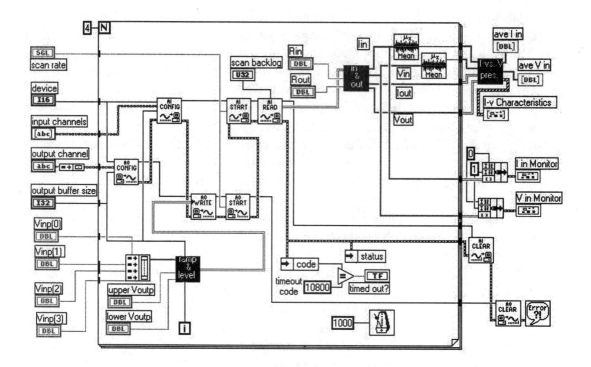

Fig. M6-8. Curve-tracer main VI diagram

Figure M6-8 shows the implementation of this curve-tracer VI in the diagram. In the lower left corner to the diagram are the controls to set up the measurement. The inputs from the **Vinp(i)** controls make up an array. The size of this array controls the maximum index of the main For Loop. These selected values for the applied input voltages, along with the applied output voltage limits from the **upper Voutp** and **lower Voutp** controls, are inputs to the **ramp & level** sub-VI inside the For Loop. This sub-VI's job is to provide the drive waveforms for the **AO Write.vi** to apply at the analog output channels 0 and 1 to the measuring circuit. The applied voltage, v_{inp}, is a DC level while the applied voltage, v_{outp}, is a ramp between the limits set by the upper **Voutp** and lower **Voutp** controls.

The **AI Read.vi** in the upper left quadrant of the For Loop reads voltages from the four analog input channels connected to the measuring circuit. These samples voltages, together with the values of the measuring resistors from the **Rin** and **Rout** controls, constitute the inputs to the **in & out** sub-VI. Its job is to use these inputs to calculate the input and output voltages and currents, v_{in}, v_{out}, i_{in}, and i_{out}, for each scan of the analog input channels of the DAQ board. The measured input voltage and current then go to the monitors on the front panel. They also get averaged by two **Mean.vi**'s before being accumulated as arrays at the edge of the For Loop. The **I vs. V pres** sub-VI has made the data ready for presentation on the front panel.

In order to keep the main VI diagram tidy, and to adhere to the top-down design approach, all the data presentation tasks have been assigned to the **I vs v pres**entation sub-VI. Its job is to prepare the outputs of the main curve-tracer VI for numerical and graphical presentation to the user.

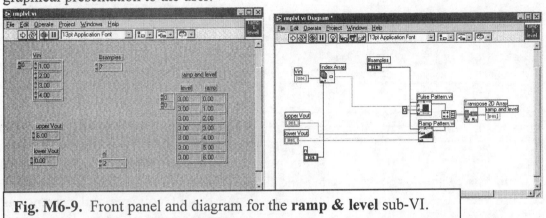

Fig. M6-9. Front panel and diagram for the **ramp & level** sub-VI.

The sub-VI's you need for the curve-tracer VI are all pretty simple. Figure M6-9 shows the front panel and the diagram for the **ramp & level** sub-VI that constructs the drive voltage waveforms for analog output. Following standard practice with sub-VI's, each input (each control on the front panel) and each output (each indicator on the front panel) is wired to a connector by right-clicking on the icon, selecting **Show Connector**, and then using the wiring tool to connect each indicator and control to a terminal on the connector. The input, **n,** of this sub-VI is wired to the index of the For Loop in the main curve-tracer VI, as shown in Fig. M6-8. In Fig. M6-9, the sub-VI has been run for a set of seven samples. Since n=2, the third element of the array, **Vin,** is selected to set the value of the **level** output at 3 volts. The seven samples constructed for the **ramp** output make up a ramp function that starts at the value from **lower Vout** and ends at the value from **upper Vout**.

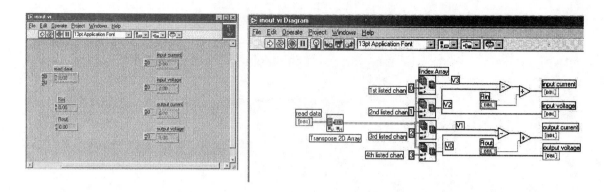

Fig. M6-10. Front panel and diagram of the **in & out** sub-VI.

Figure M6-10 shows the front panel and diagram of the **in & out** sub-VI that takes data from the **AI Read.vi** and calculates input and output voltages and currents as functions of time. The data from the **AI Read.vi** is in the form of a two dimensional array, with each row corresponding to one scan across the analog input channels, in the order they are listed in the **input channels** control on the main VI front panel. Each column of the **read data** input corresponds to the time waveform of a particular analog input channel. In the **in & out** sub-VI, each channel is stripped from the array with an **Index Array.vi**, followed by calculation of the currents using inputs from the controls, **Rin** and **Rout**. The resulting outputs are the desired input and output voltages and currents.

The For Loop in the curve-tracer VI in Fig. M6-8 accumulates two one-dimensional arrays on its right boundary:
1. The average value of i_{in} for each of the four curves traced.
2. The average value of v_{in} for each of the curves traced.

In addition, this For Loop accumulates four two-dimensional arrays on its right boundary. These arrays consist of a set of four time waveforms corresponding to each of the curves traced, one array each for the input and output voltage and current measured at the device terminals, i_{in}, v_{in}, i_{out}, and v_{out}.

The input voltage and current waveforms are used only for monitoring constant levels, so they are displayed on the small waveform graphs on the front panel of the curve tracer VI without converting the horizontal axis to time. The **I vs. v pres.vi** in Fig. M6-11 prepares the data for presentation on an x-y graph. It just passes the average values of input voltage and current to output indicators without further processing. The two-dimensional waveform arrays for output voltage and current are each separated into their four component waveforms by a sub-VI called **un-array**. At the outputs of this sub-VI, each output current is bundled with its corresponding output voltage. Then a **Build Array.vi** prepares the four *i-v* characteristic curves for plotting at the output X-Y Graph, **I vs. V**.

The **un-array** sub-VI is simple in concept, but would have been cumbersome to repeat twice in explicit form. It separates the array of four waveforms into four separate

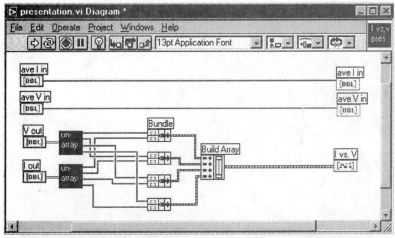

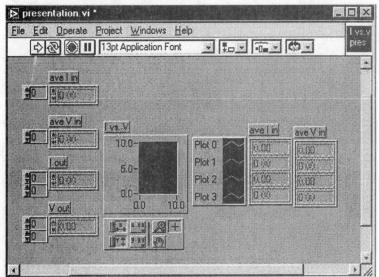

Fig. M6-11. Front panel and diagram for the **I vs.v pres** sub-VI.

waveforms, as can be seen in the example front panel in Fig. M6-12, on the following page. Its diagram shows how four **Index Array.vi**'s select rows 0-3 out of the four-row array of waveforms and deliver these individual waveforms to separate outputs.

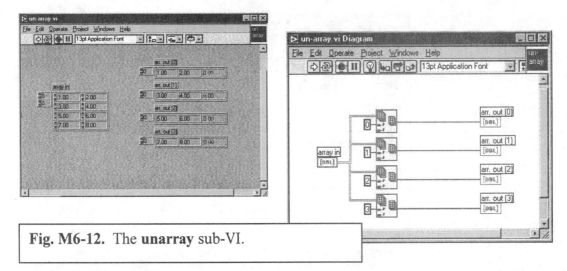

Fig. M6-12. The **unarray** sub-VI.

This curve-tracer VI and its sub-VI's constitute an excellent example of the use of *LabVIEW* sub-VI's in the top-down design of a program for a well-defined task. You can work from the top down in building these VI's. You do not need to go to the bottom level and design the **un-array** sub-VI first; instead you only need to define its inputs and outputs. You can build the front panel of a sub-VI, wire it into the main VI, and worry about the details of the sub-VI diagram later.

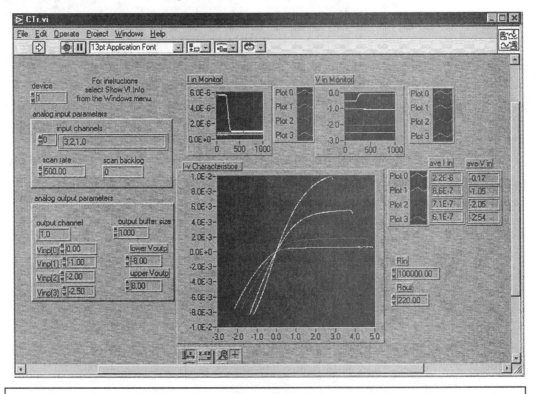

Fig. M6-13. A curve-trace showing the effects of non-constant input.

To wind up this chapter, let's explore the idea that characteristic curves for a device are families of curves with one of the input variables, either i_{in} or v_{in}, held constant within each of the curves. Since the measurement of these curves must be accomplished by applying voltages, v_{inp} and v_{outp}, to the device through two measuring resistors, there is no guarantee that either i_{in} or v_{in} will be constant during the measurement of a single characteristic curve. Figure M6-13 shows how the curve-tracer VI would respond to that situation. The device under measurement is the same JFET that was measured in Fig. M6-7. However, in the measurement shown in Fig. M6-13, the range of output voltages extends into the region where the drain-to-source voltage is negative. The **I in Monitor** and **V in Monitor** graphs show that for the first curve, Plot 0, neither i_{in} nor v_{in} is constant over the entire range of drain-to-source voltage covered in this measurement. This means that the average values of input voltage and current reported by the **ave I in** and **ave V in** indicators for Plot 0 are suspect in this measurement. Since the **I in Monitor** and **V in Monitor** graphs show that i_{in} or v_{in} are very nearly constant for Plots 1, 2 and 3, these results can be used, even though Plot 0 cannot. This curve-tracer provides more information and more flexibility in use than a conventional free standing one: in addition to plotting characteristic curves for any user-selected set of applied input voltages, it lets the user know if the input voltage and/or current are really constant over the entire output voltage range measured.

Lab Skill Exercise M6-2. Build the curve-tracer VI and its sub-VI's. Obtain a bipolar junction or a field effect transistor and generate a set of characteristic curves that cover the ranges of output voltage and output current corresponding to the *active region* of your device. Print out a copy of your VI front panel showing this measurement and attach it in the space provided.

Attach your printout from Lab Skill Exercise M6-2 here:

Chapter M6 *LabVIEW* Reference Table:			
Topic:	**Source:**		
	LabVIEW Example VI's (On-line)	*Learning with* **LabVIEW**	*LabVIEW for Everyone*
Basics	All Fundamentals categories	Chs 1 & 2	Chs. 2-5
Structures	Fundamentals-> Structures, Graphs	Ch. 5	Ch. 6
Arrays and Graphs	Fundamentals-> Arrays, Graphs	Chs. 6 & 7	Chs. 7 & 8
Waveform Generation	DAQ Analysis->Signal Generation	Ch. 8	Chs. 10 & 11
Use of Sub-VI's		Ch. 4	Chs. 3 & 4
Waveform Acquisition	DAQ	Ch. 8	Ch. 11
Spectrum Analysis	Fundamentals-> Analysis->Signal and Spectrum Analyzers	Ch. 10	
Simultaneous Analog Input/Output	DAQ-> Simultaneous Analog I/O		

Chapter M7: Small-signal Parameter Measurements

Introduction

Small-signal models are widely used to describe the behavior of nonlinear devices, such as bipolar (BJT's) and field-effect (FET's) transistors, in operating regions where their behavior is *approximately* linear. With a little work, you can calculate some of these parameters from curve-tracer data such as you obtained in Chapter M6. However, if the small-signal parameters of a two-port device or circuit constitute the information you need, in would be advantageous to measure them directly. In this chapter, you will see how to build a *LabVIEW* VI that controls a simple measuring circuit that lets you make direct measurements of any set of small-signal parameters you want.

First, let's summarize the definitions of small signal voltages and currents for a two-port circuit or device, having one port or terminal pair designated as the input, and the other designated as the output. For either port, the current is defined as positive into the device at the positive terminal. The overall *input* voltage and current waveforms necessary for a small-signal model of the device or circuit are

$$V_{in}(t) = V_{inQ} + v_{in}(t),$$
$$I_{in}(t) = I_{inQ} + i_{in}(t)$$
(M7-1)

where the quantities with the subscript, Q, are DC levels and do not vary with time. The quantities in lower-case letters are small variations away from the "quiescent" point of operation of the device (often called the "Q-point"). The *output* voltage and current waveforms are defined in exactly the same way, as

$$V_{out}(t) = V_{outQ} + v_{out}(t),$$
$$I_{out}(t) = I_{outQ} + i_{out}(t)$$
(M7-2)

If the time-varying voltages and currents in Eqs. (M7-1-2) are small enough, the characteristic curves in the neighborhood of the Q-point can be approximated by straight lines. In fact, if you agree to measure all voltages and currents relative to their Q-point values and always operate within the neighborhood of that Q-point, you can model the device or circuit with one or more of four simple, linear small-signal models:

1. The Admittance model. If you write the linear equations relating the small-signal input and output variables in terms of input and output current, they look like

$$i_{in} = y_{11}v_{in} + y_{12}v_{out}$$
$$i_{out} = y_{21}v_{in} + y_{22}v_{out}$$
(M7-3)

where the y_{ij} are the small-signal admittance parameters of the device. Of course, since the slopes and intercepts of linear approximations to the characteristic curves for a device change as you move the Q-point around, these parameters depend on the Q-point as well as on the device.

The admittance model is the one used to describe the small-signal parameters of a FET in the common source configuration. For most oxide-gate FET's and some junction FET's, y_{11} and y_{12} are both very small, since negligible current flows into or out of the gate. In the saturation region of operation, the output circuit of the FET behaves almost like a current source whose value is controlled by the input voltage, and y_{22} is small. The units of all the y-parameters are ohm^{-1} (mho).

2. The h-parameter model. This model is a popular way to describe the behavior of a common-emitter BJT in the "active" region of operation, where the emitter-base junction is forward-biased and the base-collector junction is reverse-biased. In this model, the equations are written in terms of the input voltage and the output current, as

$$v_{in} = h_{11}i_{in} + h_{12}v_{out}$$
$$i_{out} = h_{21}i_{in} + h_{22}v_{out}$$
(M7-4)

For bipolar junction transistors, h_{21} is the small-signal current amplification factor, often called the β. In this model, the units of the h-parameters are mixed: h_{21} and h_{12} are dimensionless, h_{11} is in ohms, and h_{22} is in mhos.

3. The Impedance model. This model results from inverting the matrix of y-parameters in Eq. (M7-3). The result is

$$v_{in} = z_{11}i_{in} + z_{12}i_{out}$$
$$v_{out} = z_{21}i_{in} + z_{22}i_{out}$$
(M7-5)

The units of the z-parameters are ohms.

4. The g-parameter model. This is another hybrid model that results from inverting the matrix of h-parameters in Eq. (M7-4) to give

$$i_{in} = g_{11}v_{in} + g_{12}i_{out}$$
$$v_{out} = g_{21}v_{in} + g_{22}i_{out}$$
(M7-6)

The units of the g-parameters are the inverse of those for the h-parameters.

As you might well suspect, you can convert from any one of these models to any other by manipulating the equations that define each of them, solving for different variables. Therefore, if you measure the parameters of one model, you can calculate the parameters for all the models.

It is important to remember that the parameters of these models depend strongly on the Q-point at which you choose to measure them. Your electronics textbook has descriptions of small-signal models for bipolar and field effect transistors, and also the operating regions in which these models apply. These models are usually in the chapters devoted to the specific devices. At this point, reviewing the small-signal models for the common-emitter BJT and the common-source FET would be well worth your time.

M7.1. Measuring small-signal parameters

Since you need to measure both voltage and current at the input and output sides of a two-port network to obtain the small-signal parameters, you might well expect the circuit for this measurement to be similar to the curve-tracing circuit in Chapter M6, which has to do the same thing. The measurement circuit you will use is shown in Fig. M7-1. As in the curve-tracer measurement, you will calculate the currents from measured voltage drops across measuring resistors of known value. Since the parameters you want to measure are defined by Eqs. (3-6), which relate the small-signal parts of Eqs. (M7-1,2), you will need to distinguish the small-signal variation in each measurement from its corresponding DC level. The easiest way to make that distinction is in the frequency domain, using amplitude spectra of the voltages and currents being measured: the Q-point will set their DC levels (the zero-frequency components in the amplitude spectra) while the other frequency components will determine the small-signal

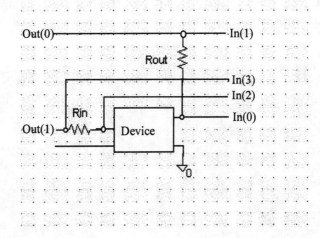

Fig. M7-1. A measurement circuit for small-signal parameters using the frequency domain to separate the small-signal components.

parameters.

To see how this measurement works, assume the voltages applied at Channel Out(1) and Channel Out(0) in Fig. M7-1 to be

$$V_3(t) = V_{3Q} + v_0 \sin(2\pi f_{in} t)$$
$$V_0(t) = V_{0Q} + v_0 \sin(2\pi f_{out} t)$$
(M7-7)

Each of these signals consists of a DC level and a small ac perturbation, and the small-signal variations are at different frequencies. Now, examine how these applied voltages propagate through the measuring circuit. Using y-parameters to model the circuit, the small-signal voltages and currents are related by

$$\frac{v_{ch3} - v_{ch2}}{R_1} = y_{11} v_{ch2} + y_{12} v_{ch1}$$
$$\frac{v_{ch0} - v_{ch1}}{R_2} = y_{21} v_{ch2} + y_{22} v_{ch1}$$
(M7-8)

These equations can be rearranged to give

$$v_{ch3} = (R_1 y_{11} + 1) v_{ch2} + R_1 y_{12} v_{ch1}$$
$$v_{ch0} = R_2 y_{21} v_{ch2} + (1 + R_2 y_{22}) v_{ch1}$$
(M7-9)

The drive voltages on the input and output sides in Fig. M7-1 insure that the small-signal voltages at channels 3 and 0 respectively contain only the frequencies, f_{in} and f_{out}. However, the voltages at channels 2 and 1 will each contain components at both frequencies. Substituting Eq. (M7-7) into Eq. (M7-9), and equating like frequencies leads to four equations, a sufficient number to solve for all four y-parameters:

at f_{in} :
$$v_{ch3}(f_{in}) = v_0 = (1 + R_1 y_{11}) v_{ch2}(f_{in}) + R_1 y_{12} v_{ch1}(f_{in})$$
$$v_{ch0}(f_{in}) = 0 = R_2 y_{21} v_{ch2}(f_{in}) + (1 + R_2 y_{22}) v_{ch1}(f_{in})$$
(M7-10)

at f_{out}:
$$v_{ch3}(f_{out}) = 0 = (1 + R_1 y_{11})v_{ch2}(f_{out}) + R_1 y_{12} v_{ch1}(f_{out})$$
$$v_{ch0}(f_{out}) = v_0 = R_2 y_{21} v_{ch2}(f_{out}) + (1 + R_2 y_{22})v_{ch1}(f_{out})$$
(M7-11)

The upper equations in Eqs. (M7-10) and (M7-11) make a pair which can be solved for y_{11} and y_{12}, while the lower equations make a pair which can be solved for y_{21} and y_{22}. The algebra involved is very tedious, but you can have *LabVIEW* do it. You can build a VI to apply the voltages at input and output, separate the two frequencies using the **Amplitude Spectrum.vi**, solve the above equations for the small-signal parameters, and display the results.

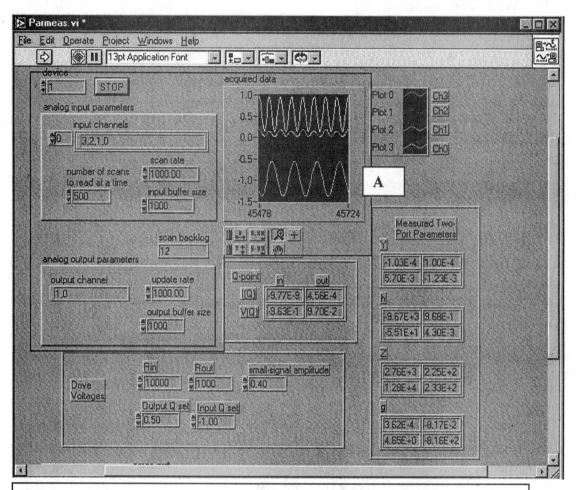

Fig. M7-2. Front panel of a VI for measuring small-signal parameters.

Figure M7-2 shows how you might start to design this VI from the top down using the front panel of the main VI as a tool in your design. The starting point is the simultaneous output/input VI from the **Examples\daq\anlog_io** library that is matched to the DAQ board in your computer. The controls and indicators inside solid line A in Fig. M7-2 are already there in the example VI.

The acquired data waveform graph shows the four voltages acquired. Channels 3 and 0 should be pure sine waves at single frequencies, as they appear here in the upper and lower traces. Channels 2 and 1 will in general contain both frequencies, as shown in the smaller-amplitude waveform in Fig. M7-2. You may or may not be able to see both waveforms for Channels 2 and 1. The reason you only see a single trace in Fig. M7-2 is that the input current in this measurement is negligible, which leads to equal voltages on Channels 2 and 1. In Fig. M7-2, the traces for Channels 2 and 1 lie right on top of each other.

The box of controls and indicators at the lower left of the front panel in Fig. M7-2, labeled **Drive Voltages**, controls the DC levels provided by the voltage sources to Channels 3 and 0 in Fig. M7-1 by means of the controls **Input Q set** and **Output Q set** respectively. These are not the Q-point DC levels for the circuit being measured, because of the additional voltage drops across R_1 and R_2 in Fig. M7-1. The actual DC voltages and currents at the input and output sides of the circuit cannot be set and must be measured, and are reported to the user in the box labeled **Q-point** in the center of the front panel. The control labeled **small-signal amplitude** sets the amplitude of the two small-signal voltages applied at Channels 3 and 0. Finally, after the VI calculates the small-signal parameters measured, it displays the results in matrix form in the box labeled Measured **Two-Port Parameters** at the lower left of the front panel. The parameters are arranged as they appear in Eqs. (M7-3)-(M7-6).

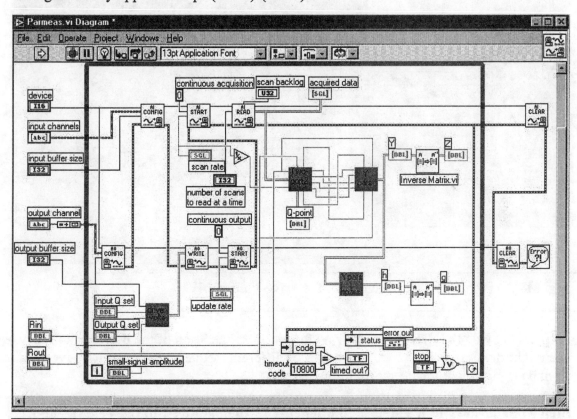

Fig. M7-3. Diagram of the parameter measurement main VI.

Setting up the front panel first, before doing any wiring, forces you to think through exactly what you want to accomplish with this VI, and what input you will need to give it. Now, let's turn to the diagram. Figure M7-3 shows the final diagram of the parameter measurement VI.

A step-by-step description of its design follows:

Step 1: Make the VI run continuously. Since you only have control of the voltage sources in Fig. M7-1, not the Q-point voltages and currents, you will have to continuously adjust the **Input Q set** and **Output Q set** controls until the indicators in the Q-point box tell you that you've reached the Q-point about which to take the measurement. To do that, your VI must run continuously, while letting you adjust the controls in the **Drive Voltages** box. To allow this continuous adjustment, remove the **While Loop** from your starting Example VI, and redraw a new one to include the **AI and AO Config.vi**'s at the left as shown in Fig. M7-3. The **Input Q set, Output Q set,** and **small-signal amplitude** controls must be inside this While Loop to make them continuously adjustable. These controls are the inputs to the **drive-volts** sub-VI in the lower left corner of the new While Loop in Fig. M7-3.

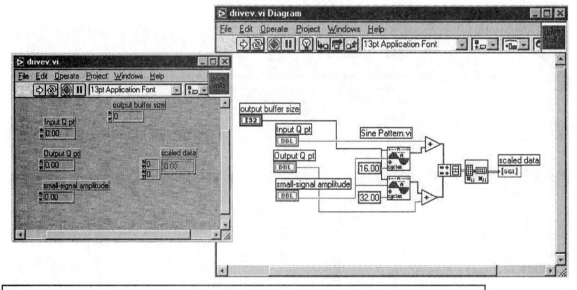

Fig. M7-4. Front panel and diagram of the **drive-volts** sub-VI.

Step 2: Provide the drive voltage waveforms. The operation of the **drive-volts** sub-VI is straightforward. It adds two sine waves of different frequencies to two DC levels. Its front panel and diagram are shown in Fig. M7-4.

Step 3: Separate the frequency components in the acquired data. The **Twoport2** sub-VI, whose front panel is shown in Fig. M7-5, takes the acquired voltage waveform data (shown in the Waveform Graph on the main VI front panel in Fig. M7-2), the input and output measuring resistor values you supply, and the sampling interval dt, and uses them to determine the DC voltage and current levels that specify the Q-point as well as the input (i) and output (o) components of the voltages, v_1 and v_2 in Fig M7-1. Figure M7-6 shows the diagram of the **Twoport2** sub-VI. The four **Index Array.vi**'s at the left separate the transposed data channels. The index constants for these VI's refer to the order of the list of **analog input channels** on the front panel of the main VI. Each channel (they are identified by Channel number in Fig. M7-6) is wired to the input of an **Amplitude and Phase Spectrum.vi**. The subtraction and division VI's calculate the

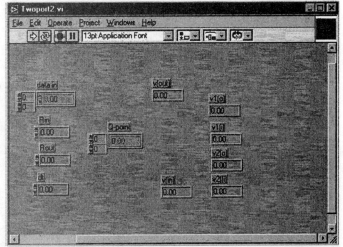

Fig. M7-5. Twoport2 sub-VI front panel.

input and output voltage and current waveforms, which then go to the bottom of the diagram to be averaged by four **Mean and Standard Deviation.vi**'s. These averages are the DC levels that make up the Q-point data.

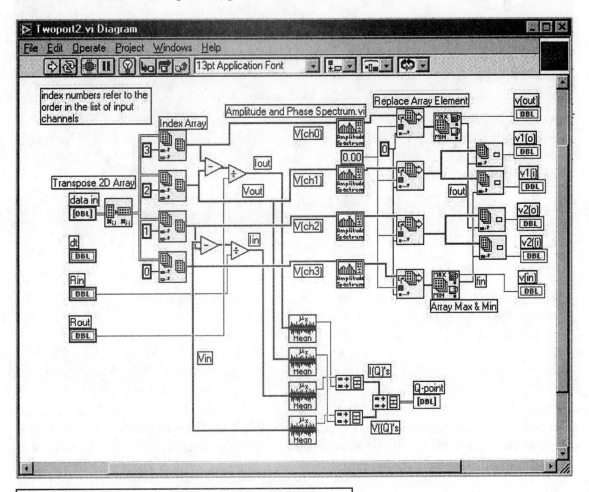

Fig. M7-6. Diagram of the **Twoport2** sub-VI.

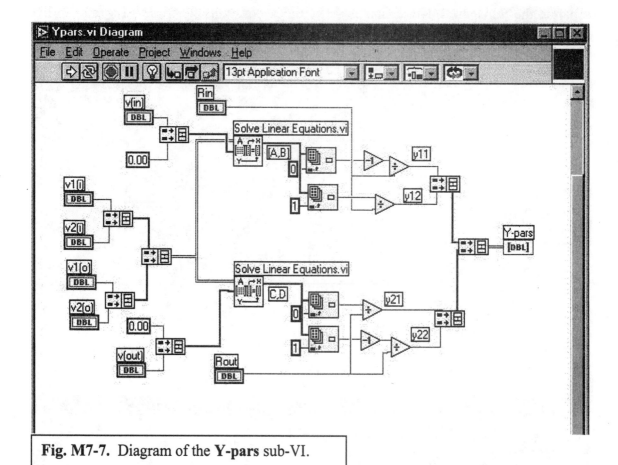

Fig. M7-7. Diagram of the **Y-pars** sub-VI.

Step 4: Compute the y-parameters from the frequency components. Figure M7-7 shows the diagram of the **Y-pars** sub-VI that solves Eqs. (M7-10) and (M7-11) for the y-parameters. The sub-VI starts out by separating Eqs. (M7-10) and (M7-11) into two pairs of equations in the form

$$\begin{bmatrix} v_0 \\ 0 \end{bmatrix} = \begin{bmatrix} v_{ch2}(f_{in}) & v_{ch1}(f_{in}) \\ v_{ch2}(f_{out}) & v_{ch1}(f_{out}) \end{bmatrix} \begin{bmatrix} A \\ B \end{bmatrix}, \quad \text{(M7-12)}$$

and,

$$\begin{bmatrix} 0 \\ v_0 \end{bmatrix} = \begin{bmatrix} v_{ch2}(f_{in}) & v_{ch1}(f_{in}) \\ v_{ch2}(f_{out}) & v_{ch1}(f_{out}) \end{bmatrix} \begin{bmatrix} C \\ D \end{bmatrix}, \quad \text{(M7-13)}$$

where we defined

$$\begin{aligned} A &= 1 + R_1 y_{11} \\ B &= R_1 y_{12} \\ C &= R_2 y_{21} \\ D &= 1 + R_2 y_{22} \end{aligned} \quad \text{(M7-14)}$$

The **Solve Linear Equations.vi**'s in Fig. M7-7 solve Eqs. (M7-12) and (M7-13) for two solution vectors [A, B] and [C,D] shown is their outputs. The VI's that follow just solve Eq. (M7-14) for the y-parameters and form them into a 2x2 array for output.

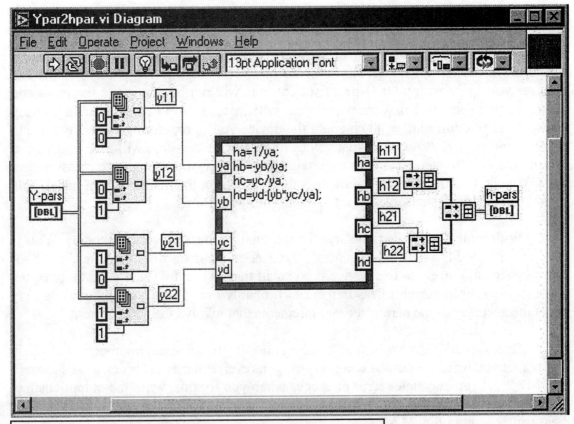

Fig. M7-8. Diagram of the **Ypar2hpar** sub-VI.

Step 5: Convert the y-parameters into the other types. Inverting the matrix of y-parameters gives the z-parameters directly. The **Ypar2hpar** sub-VI in Fig. M7-3 converts the y-parameters into h-parameters, which are then matrix-inverted to give the g-parameters. The **Ypar2hpar** sub-VI diagram is shown in Fig. M7-8. The Formula Node in the middle of the diagram implements a solution for the h-parameters in terms of the y-parameters. Formula nodes only allow two characters to represent each variable inside them. Consequently, the suffix "a" goes with the subscript 11, the suffix "b" with the subscript 12, the suffix "c" with the subscript 21, and the suffix d with the subscript 22. Thus, the formula node implements the equations,

$$h_{11} = 1/y_{11}$$
$$h_{12} = -y_{12}/y_{11}$$
$$h_{21} = y_{21}/y_{11}$$
$$h_{22} = y_{22} - y_{12}y_{21}/y_{11}$$
(M7-15)

You can prove Eq. (M7-14) by solving Eq. (M7-3) for v_{in} and i_{out}. The g-parameters follow from inverting the matrix for the h-parameters.

M7.2. Operating the parameter-measurement VI

The VI described above at considerable length operates on the same principle as the curve-tracer VI in Chapter M6. It uses two known resistors to measure current. This means that, in order to get an accurate current measurement, you must choose these resistors large enough so that the currents you expect will cause a voltage drop large enough to measure with your DAQ board. Even the sinusoidal amplitudes provided by the **small-signal amplitude** control should produce readable voltage drops in your measuring resistors. However, if you make these resistors too large, most of the DC voltages you apply with your **Q-point set** controls will not be applied to the input and output of your device being measured, but will instead drop across your measuring resistors. If you don't know much about the device you're measuring, you may have to use trial and error to determine what values of measuring resistors will give you a good measurement. You can tell if your acquired data will give you an accurate measurement by looking at it on the **acquired data** waveform graph on the main VI front panel. You should be able to see sinusoidal variation in all four waveforms.

With some devices, currents may be too small to measure very accurately. This is the case with the FET we used to generate the data on the main front panel in Fig. M7-3. The gate (input) current to this device is so small that v_{ch3} and v_{ch2} are essentially equal. The resulting input current measurement is probably not very accurate, and this may affect the accuracy of the parameter measurements that involve the input current.

One warning before you start building this VI: In some environments, *LabVIEW* will not let you write the **output channel** string in reverse numerical order, as is shown in Fig. M7-2. If you encounter error messages when you try this, write the output channel string in ascending order, and just interchange the wires for Out(0) and Out(1) in your measuring circuit in Fig. M7-2.

Lab Skill Exercise M7-1. Build the parameter-measurement VI and measure the h-parameters of a BJT biased in the middle of the active region. Use the curve-tracer VI from Chapter M6, or a free-standing curve-tracer in your laboratory, to determine the voltages and currents you want for your Q-point. Print out the front panel from the curve-tracer and label the Q-point on it. Attach it in the space provided. Then measure the small-signal parameters using the VI from this chapter. Attach a printout of its front panel showing your measurements in the space provided.

Attach your curve-tracer printout here:

Attach your small-signal measurement printout here:

Topic	Chapter M7 *LabVIEW* Reference Table: Source		
	LabVIEW Example VI's (On-line)	*Learning with* **LabVIEW**	*LabVIEW for Everyone*
Basics	All Fundamentals categories	Chs 1 & 2	Chs. 2-5
Structures	Fundamentals-> Structures, Graphs	Ch. 5	Ch. 6
Arrays and Graphs	Fundamentals-> Arrays, Graphs	Chs. 6 & 7	Chs. 7 & 8
Waveform Generation	DAQ Analysis->Signal Generation	Ch. 8	Chs. 10 & 11
Use of Sub-VI's		Ch. 4	Chs. 3 & 4
Waveform Acquisition	DAQ	Ch. 8	Ch. 11
Spectrum Analysis	Fundamentals-> Analysis->Signal and Spectrum Analyzers	Ch. 10	
Simultaneous Analog Input/Output	DAQ-> Simultaneous Analog I/O		
Linear Algebra (Matrix Operations)	Fundamentals-> Analysis-> Linear Algebra		

Project 1: Non-inverting Amplifier

Introduction:
The non-inverting amplifier employing a single op-amp is perhaps the easiest to understand of all the amplifier circuits you will encounter. The basic circuit is shown in Fig. P1-1. In Fig. P1-1, VCC is the positive supply voltage, VEE is the negative supply voltage, and V1 is the input voltage source.

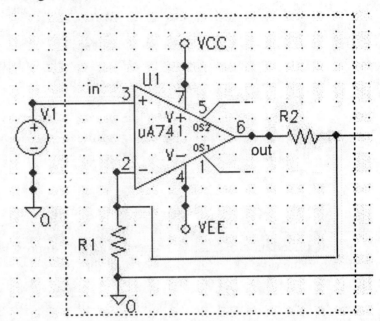

Fig. P1-1. Basic non-inverting amplifier circuit.

Summary of amplifier properties to be measured
The table below lists the specific properties of this circuit you will be measuring in this project, including some sources to consult in order to review your theoretical background on each property.

Reference Table for Project 1

Property	Hambley	Sedra & Smith	Nilsson & Riedel
Gain and phase shift vs. frequency	Sections 1.10, 2.6	Section 2.7	Sections 14.1, 14.2
Input impedance	Secton 2.4	Section 2.5	Section 5.7
Linearity	Section 2.7	Section 2.9	Section 5.2
Slew rate	Section 2.7	Section 2.9	

A very brief summary of these properties follows:
 • **Gain at low frequencies:** The output voltage in Fig. P1-1 is divided (multiplied by a ratio R1/[R1+R2]) and fed back to the inverting input. In order to keep the potential

difference between the two inputs at zero, the amplifier must supply an output voltage given by

$$v_{out} = v_{in}\left(\frac{R_1 + R_2}{R_1}\right) \quad . \tag{P1-1}$$

Thus, your choice of two resistors, R1 and R2 in Fig. P1-1, sets the low-frequency gain of the non-inverting amplifier.

• **Gain and phase shift vs. frequency:** Because of the finite gain-bandwidth product of the op-amp, this low-frequency gain will persist only up to some upper frequency limit and then decrease with increasing frequency. At frequencies where the gain is close to its low-frequency value, the phase shift of the non-inverting amplifier will approach zero. At higher frequencies, the phase shift of the output signal will differ significantly from zero. For a given op-amp, the higher the designed low-frequency gain, the lower the frequency at which gain reductions and phase shifts start to occur.

• **Input impedance:** Ideally, the input impedance of the non-inverting amplifier is infinite. This is a highly desirable feature in an amplifier, and it is worthwhile to determine how closely the real amplifier approximates this ideal behavior.

• **Output impedance:** The non-inverting amplifier is a voltage-controlled voltage source. Any real voltage source is limited in the output current it can supply to a load. The output impedance of such a source is usually represented as an additional series impedance in the output circuit, in series with the load.

• **Linearity:** Generally, your non-inverting amplifier will exhibit its ideal low-frequency gain value only over a limited range of output voltages, approximately between the positive and negative supply voltages (positive and negative "rails"), VCC and VEE in Fig. P1-1. If the amplifier input would lead to an output outside these limits, the output will "saturate" at a level near the closest rail voltage.

Measuring these amplifier properties will draw on the lab measurement skills listed in the following table:

Lab Skills Table for Project 1

Tasks	Lab Skills	Chapter
1. Simulation	AC sweep in PSPICE	Appendix I
2 & 3. Gain and phase shift vs. frequency	Complex Transfer Function Measurement	M4
4. Input impedance	Impedance Measurement	M5
5. Linearity	Waveform Capture	M2
	Frequency Spectra	M3
6. Slew rate	Waveform Capture	M2

Tasks
Task 1.1. PSPICE simulation

1.1.1. Choose values for R1 and R2 in Fig. P1-1 that would give you an amplifier gain of *approximately* 8. Choose standard resistor values that you can obtain in the lab. Enter your values, and your predicted low-frequency gain, G, here:

R1 = _____ R2 = _____ G = _____

Simulate your circuit in the frequency domain using the AC Sweep function of PSPICE. Obtain a plot of the magnitude and phase of the ratio v_{out}/v_{in}, over the range of frequency

$$50\,Hz \leq f \leq 500\,kHz \quad,$$

using a decade scale for the frequency (horizontal) axis.
- Use PROBE to create a single plot with two separate y-axes, one for the magnitude and one for the phase.
- Use the Cursor Tool in PROBE to determine the low frequency gain (at 50 Hz)
- Use the Cursor Tool in PROBE to determine the following frequencies from your simulation and enter these results in the **Low Gain** column of the table:
  - f_G, the frequency at which the gain (magnitude of v_{out}/v_{in}) begins to differ visibly from its value at the lowest frequency simulated.
  - f_{45}, the frequency at which the phase of v_{out}/v_{in} differs from zero by 45 deg.

Enter your results here:

Quantity	Low Gain	Higher Gain
Low-frequency gain (50 Hz)		
f_G		
f_{45}		
GBW product (f_G)		
GBW product (f_{45})		

- Calculate the gain-bandwidth (GBW) product using first f_G, and then f_{45}, as the value for the bandwidth and enter the results in the Low Gain column of the table.

1.1.2. Choose **new** values for R1 and R2 in Fig. P1-1 that would give you an amplifier gain of *approximately* 80. Choose standard resistor values that you can obtain in the lab. Enter your values, and your predicted low-frequency gain here:

R1 = _____ R2 = _____ G = _____

- Simulate your circuit in the frequency domain using the AC Sweep function of PSPICE. Obtain a plot of the magnitude and phase of the ratio Vout/Vin, over the range of frequency

$$50\,Hz \leq f \leq 50\,kHz \quad,$$

using a decade scale for the frequency (horizontal) axis.

- Use PROBE to create a single plot with two separate y-axes, one for the magnitude and one for the phase.
- Use the Cursor Tool in PROBE to complete the **Higher Gain** column of the table in Task 1.1.1.

Print out a copy of your first (**Low Gain**) plot from PROBE and attach it here.

Print out a copy of your second (**Higher Gain**) plot from PROBE and attach it here.

Task 1. 2. Gain and phase shift versus frequency measurement

1.2.1. Build a non-inverting amplifier circuit like the one you have designed for a gain of about 8, based on in Fig. P1-1 and your chosen values for the resistances R1 and R2. Use a 741 or similar op-amp. Using the DC power supplies on your bench, provide ± 10 volts to the VCC and VEE terminals of the op-amp. *Check the dc supply polarity to the op-amp carefully - reversing the polarity can destroy the op-amp.*

1.2.2 The gain and the phase shift of an amplifier like this one constitute respectively the magnitude and the phase of its transfer function. Consult your lab instructor and select a measurement option from the table below. If necessary, review the appropriate sections from the Measurement Chapters of this workbook.

	Measurement Option for Task 2	Workbook Sections
1	*Virtual Bench* Oscilloscope and Function Generator	M4.1
2	*LabVIEW* SFTF/F-Domain.vi (Virtual Vector Voltmeter)	M4.2.1
3	*LabVIEW* TransFunc.vi (automatic plotting)	M4.2.2

1.2.3. Draw a diagram of your measurement circuit in the space below. In addition to the components on your circuit board, be sure to show the following items on your drawing:
- All connections to the analog input and analog output channels of your DAQ board, labeled by number.
- All connections to any external devices (DC power supplies, external function generators, etc.)

Draw your measurement circuit here:

Check and adjust the magnitude of the sine wave at the input to your amplifier, and make sure it is small enough so the amplifier output does not reach the positive or negative supply voltage. Measure the gain and phase shift of your amplifier at the frequencies listed in the Data Table below, and fill in the values.

Data Table for Task 2.

Frequency (Hz)	Amplitude (volts)		Gain	Phase (degrees)
	v(in)	v(out)		
50				
100				
300				
1000				
3000				
10000				
50000				

The following notes are specific to each measurement option:
- **Option 1.** *Virtual Bench* **Oscilloscope and Function Generator.** For each frequency in the table, display v(in) and v(out) on the scope screen. Use the Cursors to compare amplitudes and to measure the phase shift between the sine waves. Fill in the amplitudes and phase shifts for each frequency in the Data Table.
- **Option 2.** *LabVIEW* **SFTF.vi** Build the **SFTF.vi**. Use it to measure amplitude ratio and phase shift. Fill out the Gain and Phase columns of the Data Table.
- **Option 3.** *LabVIEW* **TransFunc.vi.** Build the **TransFunc.vi**. Use it to get plots of the magnitude and phase of the transfer function of your amplifier. Create a cursor on both the magnitude and phase graphs of this VI and use it to read the values you need to complete the Gain and Phase columns of the Data Table.

When you have completed the table, plot the points for gain and phase shift on your PROBE plot. Use dots for gain and asterisks (*) for phase shift. Connect the data points with smooth curves.

Task 1.3. Gain vs. bandwidth trade-off.
    1.3.1. Change the resistors in your circuit to correspond with your design for higher gain. Repeat Task 2 for this amplifier. If more than one of Options 1-3 from Task 1.2 is available in your lab, it is suggested that you use a different measurement method this time.

Fill in the table below for your higher gain amplifier.

Data Table for Task 3.

Frequency (Hz)	Amplitude (volts)		Gain	Phase (degrees)
	v(in)	v(out)		
50				
100				
300				
1000				
3000				
10000				
50000				

Plot these results on your PSPICE simulation for the higher gain amplifier. Use dots for the gain and asterisks for the phase shift. Connect the symbols with smooth curves.

1.3.2. In Task 1.1.1, you completed a table comparing gain-bandwidth (GBW) products predicted by PSPICE simulation. Complete the same table below, reading the values from your graphs of experimental results:

Quantity	Low Gain	Higher Gain
Low-frequency gain (50 Hz)		
f_G		
f_{45}		
GBW product (f_G)		
GBW product (f_{45})		

Task 1.4. Input impedance measurements.

Measure the input impedance of your **higher-gain** amplifier over the same frequency range you measured the gain and phase shift. Consult your lab instructor about availability of the different measurement options below. Study the material associated with your selected option. Remember that you are trying to measure a very high impedance, so your measuring resistor will need to be very large.

Measurement Option for Task 1.4	Workbook Sections
1 *Virtual Bench* Oscilloscope	M5.1
2 *LabVIEW* **SFTF.vi** (Virtual Vector Voltmeter)	M5.2

Circle the measurement option you are using.

1.4.1. Draw your circuit for this measurement here. Label all the component values and label by channel all connections to instruments on your lab bench.

P1-10

- Enter your data in the Data Table below.

Data Table for Task 1.4.2.

Frequency (Hz)	Amplitude (volts)		Measuring Resistance (Ω)	Input Impedance Magnitude	Phase (degrees)
	Voltage waveform	Current waveform			
50					
100					
300					
1000					
3000					
10000					
50000					

Task 1.5. Linearity

1.5.1. Linearity in the time-domain. In the time domain, the onset of nonlinearity in an amplifier takes the form of distortion of the output waveform; i.e., the output ceases to be a scaled version of the input. Connect a sine wave input of adjustable amplitude (use the *Virtual Bench* Function Generator or the **FuncGen.vi** from Section M1.3) to the input of your high-gain amplifier. Connect a waveform display tool (either the *Virtual Bench* Oscilloscope or the *LabVIEW* **Acquire N Scans Analog Software Trigger.vi**) to the amplifier output. Make sure your input sine wave has zero dc offset. Gradually increase the amplitude of the input sine wave until you observe significant distortion in the output. This distortion should take the form of chopping off of the top and bottom of the output sinusoid. Display 2-3 cycles of the chopped output. Print out a copy of your display and attach it below.

Attach distorted time waveform here:

1.5.2. Linearity in the frequency domain: In the frequency domain, the onset of nonlinearity can be seen as the appearance of multiple output frequencies for a single-frequency input. Display your output waveform from Task 5.1 on the *Virtual Bench* Digital Spectrum Analyzer or on the *LabVIEW* **SpectrumAnalyzer.vi** from Section M3.3. Configure your virtual instrument so that:

- the time domain display contains almost exactly four periods of the output waveform. This will require a combination of sampling frequency, f_s, and number of samples, N, such that

$$N\left(\frac{1}{f_s}\right) = 4\left(\frac{1}{f}\right),$$

where f is the frequency of the input sine wave.
- the sampling frequency is at least 20 times the sine wave frequency.

Display the Amplitude Spectrum of the output waveform using a linear scale for the vertical axis. Increase the amplitude of the input sine wave until you can clearly see the third harmonic (at frequency $3f$). Print out a copy of the display showing the time waveform and the amplitude spectrum of the output under these conditions and attach it in the space provided.

Attach your printout of the time waveform and amplitude spectrum of the amplifier output from Task 1.5.2 here:

Task 1.6. Slew Rate

The slew rate of an amplifier based on an op-amp takes the form of an upper limit on the magnitude of the time derivative of the output voltage, $|dv_{out}/dt|$. This means that if your amplifier is driven with a square wave whose frequency is within the region where it exhibits maximum gain, it will output an approximation to this square wave, whose rising and falling sections will not be vertical, but rather have a slope equal to the slew rate.

Drive your amplifier from Task 1.5 with a 0.1 volt amplitude, zero dc offset square-wave of the same frequency, f, you used in Task 1.5. Display the resulting amplifier output on the *Virtual Bench* Oscilloscope, or on the *LabVIEW* **Acquire N Scans Analog Software Trigger .vi.**. Adjust the triggering so that the rising edge of the output square wave is displayed on the screen. Adjust the Time Base on the *Virtual Bench* Oscilloscope, or the sampling frequency on the *LabVIEW* **Acquire N Scans Analog Software Trigger .vi.** to make the finite slope clearly visible. Then use two cursors to measure this slope:

- If you are using the *Virtual Bench* Oscilloscope, turn on the two available cursors and place them at the top and the bottom of the rising edge of the output waveform.
- If you are using the *LabVIEW* **Acquire N Scans Analog Software Trigger .vi.**, create two new Cursors, lock them to the waveform, and then place them at the top and the bottom of the rising edge of the output waveform.

In both cases, your screen should resemble Fig. P1-2.

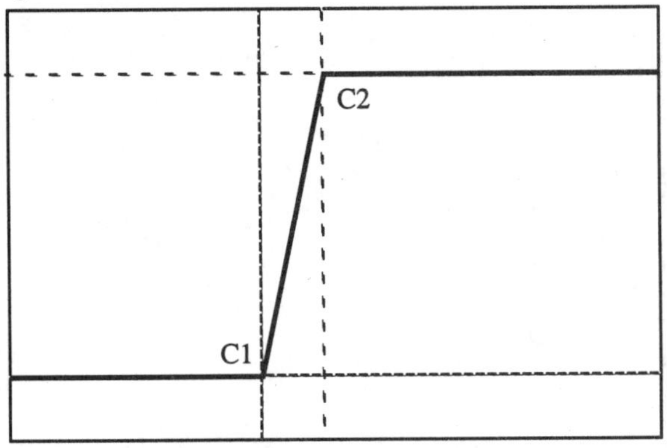

Fig. P1-2.
Waveform and Cursor configuration for measuring slew rate.

Warning: Be sure your sampling rate is high enough so that there is at least one sample point (ideally, there should be several) between cursors C1 and C2. If your sampling period is of the same order as the time it takes the output waveform to rise from the bottom to the top of the square wave, your slew rate measurement will not be accurate. Enter the data from your screen (subscripts correspond to cursor number) in the Data Table to calculate the slew rate, and attach a printout of your display screen (with the cursors in the right positions) on the following page.

Data table for Task 1.6

V_1	(v)	V_2	(v)
T_1	(___ sec)	T_2	(___ sec)
$dV=|V_1-V_2|$	(v)	$dT=|T_1-T_2|$	(___ sec)
Calculated slew rate dV/dT			(v/sec)

Attach your virtual instrument screen printout from Task 1.6 here:

Project 2: Inverting Amplifier

Introduction

The inverting amplifier employing a single op-amp is shown in Fig. P2-1. In Fig. P2-1, VCC is the positive supply voltage, VEE is the negative supply voltage, and V1 is the input voltage source.

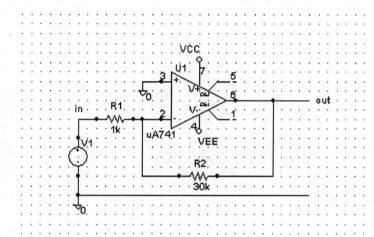

Fig. P2-1. Basic inverting amplifier circuit.

Summary of amplifier properties to be measured

The table below lists the specific properties of this circuit you will be measuring in this project, including some sources to consult in order to review your theoretical background on each property.

Reference Table for Project 2

Property	Hambley	Sedra & Smith	Nilsson & Riedel
Gain and phase shift vs. frequency	Sections 1.10, 2.6	Section 2.3	Sections 14.1, 14.2
Input impedance	Secton 2.3	Section 2.3	Section 5.7

A very brief summary of these properties follows:

• **Gain at low frequencies:** The low-frequency gain of the inverting amplifier is given by

$$\frac{v_{out}}{v_{in}} = -\frac{R_2}{R_1} \qquad \text{(P2-1)}$$

Thus, your choice of two resistors, R1 and R2 in Fig. P2-1, sets the low-frequency gain of the inverting amplifier. The inversion arises from the minus sign in Eq. (P2-1).

- **Input impedance:** The inverting amplifier does not share the infinite input impedance property of the non-inverting amplifier. Inspection of Fig. P2-1 shows that, if the op-amp is ideal, then the input impedance of the inverting amplifier is just R_1. This leads to a tradeoff in design between high gain and high input impedance. Such a tradeoff is not so evident in the non-inverting amplifier of Project 1.

- **Other properties.** Other properties of inverting amplifiers are very similar to those of inverting ones.

Measuring these amplifier properties will draw on the lab measurement skills listed in the following table:

Lab Skills Table for Project 2

Tasks	Lab Skills	Chapter
1. Simulation	AC sweep in PSPICE	Appendix I
2 & 3. Gain and phase shift vs. frequency	Complex Transfer Function Measurement	M4
4. Input impedance	Impedance Measurement	M5

Tasks
Task 2.1. PSPICE simulation

2.1.1. Choose values for R1 and R2 in Fig. P2-1 that would give you an amplifier gain of *approximately* -8. Choose standard resistor values that you can obtain in the lab. Enter your values, and your predicted low-frequency gain, G, here:

R1 = _____   R2 = _____   G = _____

Simulate your circuit in the frequency domain using the AC Sweep function of PSPICE. Obtain a plot of the magnitude and phase of the ratio v_{out}/v_{in}, over the range of frequency

$$50 Hz \leq f \leq 500 kHz \quad ,$$

using a decade scale for the frequency (horizontal) axis.
- Use PROBE to create a single plot with two separate y-axes, one for the magnitude and one for the phase.
- Use the Cursor Tool in PROBE to determine the low frequency gain (at 50 Hz)
- Use the Cursor Tool in PROBE to determine the following frequencies from your simulation and enter these results in the **Low Gain** column of the table:
  - f_G, the frequency at which the gain (magnitude of v_{out}/v_{in}) begins to differ visibly from its value at the lowest frequency simulated.
  - f_{45}, the frequency at which the phase of v_{out}/v_{in} differs from 180 deg by 45 deg.

Enter your results here:

Quantity	Low Gain	Higher Gain
Low-frequency gain (50 Hz)		
f_G		
f_{45}		
GBW product (f_G)		
GBW product (f_{45})		

• Calculate the gain-bandwidth (GBW) product using first f_G, and then f_{45}, as the value for the bandwidth and enter the results in the Low Gain column of the table.

2.1.2. Choose **new** values for R1 and R2 in Fig. P2-1 that would give you an amplifier gain of *approximately* -80. Choose standard resistor values that you can obtain in the lab. Enter your values and your predicted low-frequency gain here:

R1 = _____ R2 = _____ G = _____

• Simulate your circuit in the frequency domain using the AC Sweep function of PSPICE. Obtain a plot of the magnitude and phase of the ratio Vout/Vin, over the range of frequency

$$50 Hz \leq f \leq 500 kHz \quad ,$$

using a decade scale for the frequency (horizontal) axis.

• Use PROBE to create a single plot with two separate y-axes, one for the magnitude and one for the phase.

- Use the Cursor Tool in PROBE to complete the **Higher Gain** column of the table in Task 2.1.1.

Print out a copy of your first (**Low Gain**) plot from PROBE and attach it here.

Print out a copy of your second (**Higher Gain**) plot from PROBE and attach it here.

Task 2.2. Gain and phase shift versus frequency measurement

2.2.1 Build a non-inverting amplifier circuit like the one you have designed for a gain of about -8, based on in Fig. P2-1 and your chosen values for the resistances R1 and R2. Use a 741 or similar op-amp. Using the DC power supplies on your bench, provide ± 10 volts to the VCC and VEE terminals of the op-amp. *Check the dc supply polarity to the op-amp carefully - reversing the polarity can destroy the op-amp.*

2.2.2. The gain and the phase shift of an amplifier like this one constitute respectively the magnitude and the phase of its transfer function. Consult your lab instructor and select a measurement option from the table below. If necessary, review the appropriate sections from the Measurement Chapters of this workbook.

	Measurement Option for Task 2.2	Workbook Sections
1	*Virtual Bench* Oscilloscope and Function Generator	M4.1
2	*LabVIEW* F-Domain.vi (Virtual Vector Voltmeter)	M4.2.1
3	*LabVIEW* TransFunc.vi (automatic plotting)	M4.2.2

Draw a diagram of your measurement circuit in the space below. In addition to the components on your circuit board, be sure to show the following items on your drawing:
- All connections to the analog input and analog output channels of your DAQ board, labeled by number.
- All connections to any external devices (DC power supplies, external function generators, etc.)

Draw your measurement circuit here:

Check and adjust the magnitude of the sine wave at the input to your amplifier, and make sure it is small enough so the amplifier output does not reach the positive or negative supply voltage. Measure the gain and phase shift of your amplifier at the frequencies listed in the Data Table below and fill in the values.

Data Table for Task 2.2

Frequency (Hz)	Amplitude (volts)		Gain	Phase (degrees)
	v(in)	v(out)		
50				
100				
300				
1000				
3000				
10000				
50000				

The following notes are specific to each measurement option:
- **Option 1.** *Virtual Bench* **Oscilloscope and Function Generator.** For each frequency in the table, display v(in) and v(out) on the scope screen. Use the Cursors to compare amplitudes and to measure the phase shift between the sine waves. Fill in the amplitudes and phase shifts for each frequency in the Data Table.
- **Option 2.** *LabVIEW* **SFTF.vi.** Build the **SFTF.vi**. Use it in the Transfer Function mode to measure amplitude **ratio** and phase shift. Fill out the Gain and Phase columns of the Data Table.
- **Option 3.** *LabVIEW* **TransFunc.vi.** Build the **TransFunc.vi**. Use it to get plots of the magnitude and phase of the transfer function of your amplifier. Create a cursor on both the magnitude and phase graphs of this VI, and use it to read the values you need to complete the Gain and Phase columns of the Data Table.

When you have completed the table, plot the points for gain and phase shift on your PROBE plot. Use dots for gain and asterisks (*) for phase shift. Connect the data points with smooth curves.

Task 2.3. Gain vs. bandwidth trade-off.

2.3.1. Change the resistors in your circuit to correspond with your design for higher gain. Repeat Task 2.2 for this amplifier. If more than one of Options 1-3 from Task 2.2 is available in your lab, it is suggested that you use a different measurement method this time.

Fill in the table below for your higher gain amplifier.

Data Table for Task 2.3

Frequency (Hz)	Amplitude (volts)		Gain	Phase (degrees)
	v(in)	v(out)		
50				
100				
300				
1000				
3000				
10000				
50000				

Plot these results on your PSPICE simulation for the higher gain amplifier. Use dots for the gain and asterisks for the phase shift. Connect the symbols with smooth curves.

2.3.2. In Task 2.1.1, you completed a table comparing gain-bandwidth (GBW) products predicted by PSPICE simulation. Complete the same table below, reading the values from your graphs of experimental results.

Quantity	Low Gain	Higher Gain
Low-frequency gain (50 Hz)		
f_G		
f_{45}		
GBW product (f_G)		
GBW product (f_{45})		

Task 2.4. Input impedance measurements.

Measure the input impedance of your **higher-gain** amplifier over the same frequency range you measured the gain and phase shift. Consult your lab instructor about availability of the different measurement options below. Study the material associated with your selected option. Remember that your measuring resistor should be comparable in value to the magnitude of the impedance you want to measure.

Measurement Option for Task 2.4	Workbook Sections
1 *Virtual Bench* Oscilloscope	M5.1
2 *LabVIEW* SFZ.vi (Virtual Vector Voltmeter)	M5.2

Circle the measurement option you are using.

2.4.1. Draw your circuit for this measurement here. Label all the component values and label by channel all connections to instruments on your lab bench.

- Enter your data in the Data Table below.

Data Table for Task 2.4

Frequency (Hz)	Amplitude (volts)		Measuring Resistance (Ω)	Input Impedance Magnitude	Phase (degrees)
	Voltage waveform	Current waveform			
50					
100					
300					
1000					
3000					
10000					
50000					

Project 3: Artificial Inductor

Introduction

Unfortunately for those who want to design circuits with them, real inductors suffer from a number of drawbacks, such as bulk, expense, and high internal resistance. Once again, the ever-useful op-amp can come to the rescue, so long as the voltages and frequencies involved stay in the ranges where the op-amps behave in nearly ideal fashion.

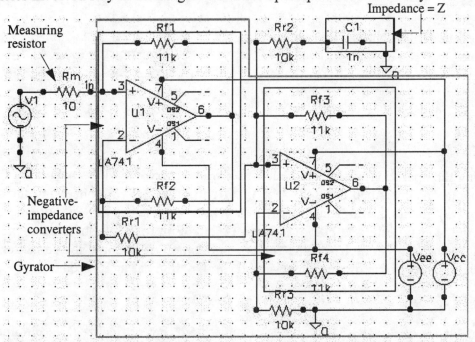

Fig. P3-1. A gyrator circuit, shown converting a capacitor, C1, into an inductance looking into the circuit from the node, **in**.

Figure P3-1 shows a circuit called a gyrator, in which two op-amps combine to make the capacitor behave like an inductor with a value of
$$L = (Rr)^2 C \qquad (P3-1)$$
In this project, you will show in a PSPICE simulation and with laboratory measurements, that the impedance of this circuit looking into the input terminals is
$$Z_{in} = j\omega (Rr)^2 C = \frac{(Rr)^2}{Z} \quad , \qquad (P3-2)$$
where Z is the impedance of the circuit inside the box at the upper right corner of Fig. P3-1.

When there is a capacitor inside this box, as shown in P3-1, the circuit looks like an inductor of the value given in Eq. (P3-1). As you can see from the right hand side of Eq. (P3-2), the gyrator can be thought of as a circuit which *inverts* the impedance to which it is connected.

The gyrator circuit in Fig. P3-1 contains two Negative Impedance Converters (NIC's), one of which is shown separately in Fig. P3-2. This circuit can be thought of as producing at its terminals the *negative* of the impedance to which it is connected.

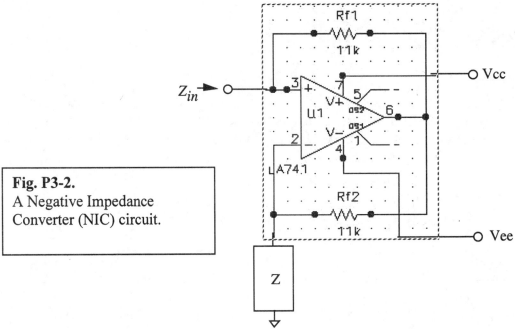

Fig. P3-2.
A Negative Impedance Converter (NIC) circuit.

Verifying the operation of the NIC is left as a preparatory exercise. Assuming that the NIC operates as described here, the second NIC in Fig. P3-1 (the one using the second op-amp, U2) inverts the impedance of Rf3 to give an impedance of –Rf3 looking into its input (the non-inverting input of op-amp U2). Figure P3-1 shows this negative impedance to be in parallel with the series combination of Rr2 and the capacitor at the upper right corner of the figure. This parallel combination has an impedance of

$$-R_r \parallel (R_r + \frac{1}{j\omega C}) = -R_r^2 j\omega C - R_r. \quad \text{(P3-3)}$$

The impedance of Eq. (P3-3) combines in series with that of Rr1 = R_r to eliminate the second term on the right side of Eq. (P3-3), before being made negative by the first NIC in Fig. P3-1 to yield the overall impedance in Eq. (P3-2).

Reference Table for Project 3

Topic	Hambly	Sedra & Smith	Nilsson & Riedel
Inductors	Section 11.6		Section 6.1
Artificial Inductor		Section 11.6	
Impedance			Section 9.4

Tasks

Task 3.0. Preparation

3.0.1. Review the reference material in the Reference Table for this Project.

3.0.2. Show that the NIC in Fig. P3-2 works as described, i.e define the voltage and current into the NIC as V_{in} and I_{in}, and the voltage and current at the impedance Z as V_z and I_z, and show that

$$\frac{V_{in}}{I_{in}} = -\frac{V_z}{I_z}.$$

Write your proof here:

Task 3.1. Computer Simulation

No matter how clever we are at modeling the behavior of a circuit, we can usually find a set of conditions for which our carefully constructed circuit model breaks down and becomes unacceptably inaccurate. Our earlier lab and simulation work with op-amps has told us to be careful about two things:

1. Driving the op-amp so that its output tries to leave the range, $-V_{ee} < V_{out} < V_{cc}$. This happens when the voltage difference between the inverting and non-inverting inputs becomes non-negligible.
2. Changing the voltage difference between the inverting and non-inverting inputs too fast for the op-amp's slew-rate limit.

PSPICE's model for the 741 op-amp accurately predicts its behavior even outside of the ranges of voltage and frequency for which its behavior is nearly ideal. Therefore, PSPICE simulations can be used to show the conditions under which a circuit design based on ideal op-amp behavior will begin to fail. In this task, you will use PSPICE to compare a "real" inductor to an artificial one based on Fig. P3-1.

3.1.1. Create a SCHEMATIC for the circuit in Fig. P3-1. Set VCC and VEE to +10 and -10 volts DC, respectively. Add a second measuring resistor (500 Ω) in series with a 100 mH inductor driven by the sinusoidal source at the input. Your schematic should look like the one shown in Fig. P3-3.
Calculate the value of the inductor "synthesized" by the gyrator circuit. Enter your result here:

L_{gyr} = _____ mH.

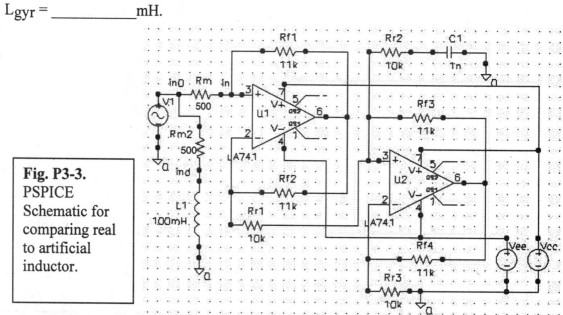

Fig. P3-3. PSPICE Schematic for comparing real to artificial inductor.

3.1.2. Use the AC Sweep analysis to obtain a PROBE plot of the **magnitudes** of the impedances looking into the real inductor (node = **ind**) and looking into the artificial inductor (node = **in**). Both curves should be on the same plot, and the horizontal axis should be a decade scale over the range 10 Hz to 1 MHz. The vertical axis should be a log scale. Mark on your plot the highest frequency for which the magnitude of the artificial inductor's impedance closely resembles that of the real one. Attach your plot in the space provided.

3.1.3. Obtain a similar plot comparing the **phases** of the impedances of the real and the artificial inductor. Use the same decade scale and frequency range for the horizontal axis. Use a linear scale (degrees) for the vertical axis. Attach the result in the space provided. Mark on your plot the highest frequency for which the phase of the artificial inductor's impedance closely resembles that of the real one.

Attach Task 3.1.2 printout here:

Attach Task 3.1.3 printout here:

3.1.4. So far, it looks as if your artificial inductor could replace a real one at all frequencies below the upper limits you found in Tasks 3.1.2-3. This next simulation will show you that PSPICE can fool you if you only use AC Sweep analyses to simulate your artificial inductor.

a) Perform a Transient analysis on your PSPICE Schematic from Task 3.1.2. As input, use a sine wave of one volt amplitude, at a frequency just below the upper frequency limits you found in Tasks 3.1.2-3. Use PROBE to display 2-3 cycles of the voltage and the current *into* the artificial inductor at the node **in**. Satisfy yourself that the voltage and current waveforms predicted by PSPICE are sine waves and that the current lags the voltage by 90 deg, as you would expect.

b) Now, repeat part a), but set the frequency of your sine wave input very low (about 10-50Hz). Notice that the waveforms are no longer sinusoidal. Print out a display of 2-3 cycles of the voltage and current waveforms. Attach it in the space provided.

What you have discovered here is another limitation of this op-amp based artificial inductor: the impedance of a real inductor approaches zero at low frequencies and therefore large currents will flow for even a very small applied voltage. The op-amps cannot supply the currents required, and the artificial inductor fails to mimic the real one if the frequency is too low.

Attach your printout from Task 3.1.4 here:

Task 3.2. Comparative impedance measurements

3.2.1. Build the artificial inductor circuit of Fig. P3-1. Use 11 k resistors for all the Rf's. Keep all Rr's equal, but use a combination of values for Rr and C1 that gives you a predicted artificial inductance somewhere in the range 200-500 mH.

3.2.2. Select a precision ($\pm 1\%$) measuring resistor of known value. **The 500 Ω value in Fig. P3-1 is all right for PSPICE simulations, but too small for measurement purposes.** The range 1K-5KΩ is about right.

Enter your chosen measuring resistor value here:

$$R_m = \underline{\hspace{3cm}} \Omega$$

3.2.3. In consultation with your lab instructor, choose a method to measure the impedance of your artificial inductor from one of the following:

1. *Virtual Bench* Oscilloscope – Section M5.1
2. *LabVIEW* Single-frequency Impedance measurement – Section M5.2
3. *LabVIEW* Broad-Band Impedance measurement – Section M5.3

Draw your measurement circuit below. You may draw the artificial inductor circuit as a box and show only its input node, **in**. However, show all other connections to analog input and output channels, power supply voltages VEE and VCC, etc.

Draw your measurement circuit here:

If you are using method 1 or 2, fill in the applicable columns of the Data Table for Task 3.2.3. For method 1, fill in all columns; for method 2, fill in the last two columns.

Data Table for Task 3.2.3: Artificial inductor measurements

Freq. (Hz)	Voltage Amplitude (v)	Math Channel Amplitude (v)	Current Amplitude (A)	$|Z|$ (Ω)	Phase of Z (deg)
30					
100					
300					
1000					
3000					
10000					
30000					
100000					

- If you used Method 1 for this task, then your waveforms for some of the lower frequencies in the Data Table will not be sine waves, and you will not be able to determine relative amplitudes and phases. Draw lines through the rows in the Data Table representing frequencies at which your current waveform is not a sine wave.
- If you used method 2 or 3, use one of the methods in Section M2.2 to display the two waveforms used as analog inputs for methods 2 and 3. If the input is a sine wave, the output should also be a sine wave. For low enough frequencies, you will observe significant distortion. Find the lower frequency limit at which you still see sine waves for the input and output.

3.2.4. Based on your results from Task 3.2.3, enter the range of frequencies over which the your artificial inductor acts similarly to an ideal one:

_____ Hz < f < _____ Hz

If you used the Broad-band impedance VI for this task, print out a copy of the front panel with your results, and attach it here:

Attach method 3 printout for Task 3.2.3 here:

Project 4: Transformers and Impedance

Introduction

Project 3 showed an example of how op-amp circuits can be used to transform impedances - - effectively converting a capacitor into an inductor. For applications where the only impedance transformation the designer wants is one of magnitude, the transformer has served well.

Transformer references

Hambly	Sedra and Smith	Nilsson and Riedel
Section 10.6	Section 3.6	Section 9.10

Transformers operate on the principle of mutual inductance, whereby magnetic coupling between two inductors causes the voltage across each to depend on the current in the other. In analysis of circuits containing transformers, it is convenient to assume the transformer is ideal; that is, that the two inductances are very large, the magnetic coupling is very strong, and that the resistances associated with the inductive coils are negligible. Many real transformers are far from ideal. This less than ideal device is usually represented within a circuit as shown in Fig. P4-1.

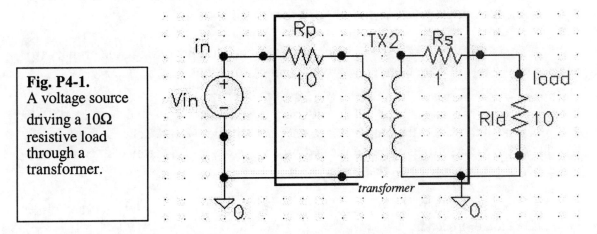

Fig. P4-1. A voltage source driving a 10Ω resistive load through a transformer.

For simulation purposes, the primary and secondary coils of the transformer are given associated resistances Rp and Rs, respectively. To model the transformer in PSPICE, you must give the inductance values for the two coils, and the coupling coefficient, as shown in Fig. P4-2.

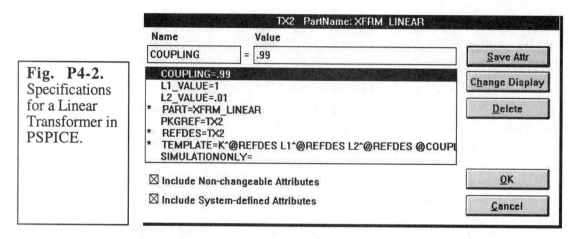

Fig. P4-2. Specifications for a Linear Transformer in PSPICE.

The mutual inductance is calculated from the quantities in Fig. P4-2 as

$$M = k\sqrt{L_1 L_2} \;, \tag{P4-1}$$

where:
  L_1 is the inductance of the primary coil,
  L_2 is the inductance of the secondary coil, and
  k is the coupling coefficient (COUPLING in Fig P4-2).

For sine-wave excitation, the voltages across and currents through the primary (p) and secondary (s) coils are then related by the pair of equations:

$$V_p = R_p I_p + j\omega L_p I_p \pm j\omega M I_s \tag{P4-2a}$$
$$V_s = R_s I_s + j\omega L_s I_s \pm j\omega M I_p. \tag{P4-2b}$$

The proper sign to use for the mutual inductance term in Eq. (P4-2) depends on how the windings in the transformer were fabricated. In lab, you will determine it empirically, since it may not be shown on a real transformer.

A look at Eqs. (P4-2) suggests how the measurements of relevant transformer properties should proceed:
  a) If the secondary is open ($I_s = 0$), then the impedance measured looking into the primary terminals is just that of the primary coil alone.
  b) If the primary is open ($I_p = 0$), then the impedance measured looking into the secondary terminals is just that of the secondary coil alone.
  It is reasonable to expect these impedances to look like those of a real inductor: i.e., a real part containing the resistance of the coil windings and an imaginary part containing the inductance of the winding.
  c) If the secondary is connected to a voltage-measuring device of very high impedance (such as the analog input terminals of a DAQ board), I_s is still approximately zero, and Eq. (P4-2b) gives the voltage measured at the secondary as

$$V_s = \pm j\omega M I_p = \pm j\omega M \left(\frac{V_p}{R_p + j\omega L_p} \right) \tag{P4-3}$$

By convention, M is a positive number, so the appropriate sign in Eq.(P4-3) is the one that gives the correct measured phase relationship between V_s and I_p.

In summary, a series of three impedance measurements can provide all the values needed to specify the properties of a linear transformer:
   a) measurement of the input impedance of the primary, with the secondary open,
   b) measurement of the input impedance of the secondary, with the primary open, and
   c) measurement of a trans-impedance, V_s / I_p, with the secondary open.

Tasks

Task 4.0. Preparation

4.0.1. Review the reference material on transformers cited above.

4.0.2. Create the schematic Fig. P4-1 in PSPICE. Perform an AC sweep using a decade scale over the range 10Hz to 100KHz. Print out a plot with two Y-axes showing
 - The magnitude of the impedance between the node **(in)** and ground
 - The phase of the impedance between the node **(in)** and ground

Attach your plot in the space provided.

4.0.3. If the circuit in Fig. P4-1 contained an **ideal** transformer, the impedance seen at the input to the primary would be just

$$Z_{in} = \frac{L_p}{L_s} Z_{load} = \frac{L_p}{L_s} R_{ld} \,. \tag{P4-4}$$

Label on your plot from Task 4.0.2 the frequency range for which Eq. (P4-4) is (very) roughly true.

Attach the printout from Task 4.0.2 here:

Task 4.1. Characterizing a transformer

4.1.1. Measure the impedances (at 200 Hz) of the two coils in your transformer, performing each measurement with the opposite coil open, as described above.

Choose a measuring resistor large enough so that the current to be measured produces a voltage drop of at least 0.1 v. Use one of the following methods:

Methods for Transformer Impedance Measurement

Virtual Bench Oscilloscope	Section M5.1
LabVIEW Single-Frequency Impedance Measurement	Section M5.2

Draw your measurement circuit here. Show all analog input and output connections.

Measurement circuit for Task 4.1.1:

4.1.2. Calculate the impedances you measured from the output of your instrument in Task 4.1.1. Fill in the appropriate columns in the Data Table.

Enter your final value for the measuring resistor: R_m = _____

Data Table for Task 4.1.2

Coil	Voltage Amplitude (v)	Math Channel Amplitude (v)	Current Amplitude (A)	$|Z|$ (Ω)	Phase of Z (deg)
1					
2					

4.1.3. Now measure the trans-impedance of your transformer, using either the *Virtual Bench* or the *LabVIEW* method described in Section M5.4. Print out your instrument front panel showing your results for this measurement and attach it in the space provided.

Attach the printout from Task 4.1.3 here:

4.1.4. From the data on your printed screen, calculate the value of *M*, the correct sign to be used in Eq. (P4-2), and the coupling coefficient *k*, for your transformer. Show your calculation below and enter the results.

M = _____ h;   Sign = _____ ; *k* = _____

Task 4.2. Simulation of the transformer under test in a circuit

4.2.1. Modify the component values in the PSPICE Schematic for Task 4.0.2 to reflect the values you measured for your transformer in Task 4.1. **Make the coil with the larger inductance the primary coil in your circuit.** Perform an AC sweep using a decade scale over the range 10Hz to 100KHz. Print out a plot with two Y-axes showing
  • The magnitude of the impedance between the node **in** and ground
  • The phase of the impedance between the node **in** and ground
Attach your plot in the space provided.

4.2.2. Label on your plot the range of frequencies for which your transformer most closely approximates an ideal one. Don't be surprised if your transformer is far from ideal.

Attach your printout from Task 4.2.1 here:

Task 4.3. Measuring the impedance conversion properties of a real transformer.

4.3.1. Assemble a circuit like the one in Fig. P4-1, using your real transformer. **Make sure you use the coil with the lower inductance value as the secondary.** Put a suitable measuring resistor of known value in series between the voltage input and the primary. Connect a 10Ω load resistor to the secondary. Use either the *Virtual Bench* Oscilloscope or the Single Frequency Impedance VI to measure the impedance looking into the primary of your transformer at a frequency where you predicted your transformer to be nearly ideal in Task 4.2.2. Attach a printout of your instrument screen in the space provided. The sine wave input to this circuit should be no larger than 2v amplitude.

4.3.2. From the data on your printed screen, calculate the magnitude and phase of the input impedance you measured. Show your calculation and enter the results below.

Measuring resistor, R_m = _____

$|Z_{in}|$ = _____   arg { Z_{in} } = _____

How does this result compare with the input impedance predicted by your simulation for Task 4.2 at this frequency?

Attach your front panel showing the results for Task 4.3.1. here:

Project 5: First- and Second-order Low Pass Filters

Introduction

Filters are two-port networks (having two terminal pairs, one each for the input, v_i and the output, v_o) which are designed to have the complex transfer function,

$$H(\omega) = \frac{V_o(\omega)}{V_i(\omega)}, \qquad (P5\text{-}1)$$

depend on frequency in some prescribed way. This project concentrates on low-pass filters, which pass low frequencies and "filter out" higher ones. Thus, we would expect $|H(\omega)|$ to be finite near $\omega = 0$ and tend toward zero at high frequencies. For some low-pass filters, the phase shift of $H(\omega)$ with frequency is not a concern, and the designer will accept whatever phase shift results, as long as $|H(\omega)|$ stays within specifications. For other filter applications, the phase shift can cause significant waveform distortion and must be dealt with.

Reference Table for Project 5

Filter Type	Hambly	Sedra & Smith	Nilsson & Riedel
1st order Low-Pass	Section 11.1	Sections 11.1, 11.4	Sections 7.2, 7.3
2nd order Low-Pass	Section 11.1	Sections 11.1, 11.4, 11.6	Sections 14.1, 14.2
Low-Pass Notch		Sections 11.1, 11.4, 11.6	

We can analyze filters in the frequency- or the time-domain, and confirm such analyses experimentally by means of frequency response or transient response measurements. There are a great many very good filter designs available. This project will only touch on a few of the simplest ones. You will be measuring the frequency and transient responses of the following filters:

First-order op-amp low-pass filter. The simplest low-pass filter is of course the series RC circuit. This simple circuit has a filter response that is too dependent on the load it is driving, so an op-amp is often used to buffer the load. Fig. P5.1 shows a single voltage source driving both a simple RC low-pass filter, and an op-amp enhanced version of the same filter. Note that in this circuit, only the op-amp filter is driving a 1KΩ load. The simple RC circuit's output is an open circuit. The frequency responses of these filters are approximately flat from dc up to some break frequency determined by the time constant of the RC network, and then decrease monotonically toward zero.

Second-order low-pass filter. Second-order low-pass filters have many possible realizations. The one in Figure P5-2 uses the artificial inductor from Project 3. The advantage of the more complex second order filter over the first-order version is that its response drops faster at higher frequencies, which leads to a sharper transition between the pass-band and the stop-band. A potential disadvantage is a peak in the frequency response at the high end of the pass-band.

Low-pass notch filter. For some applications, the designer may want the filter transfer function to drop all the way to zero at some frequency outside the pass-band.

The low-pass notch does this by placing a second capacitor in parallel with the (artificial) inductor in the second-order low-pass filter described above.

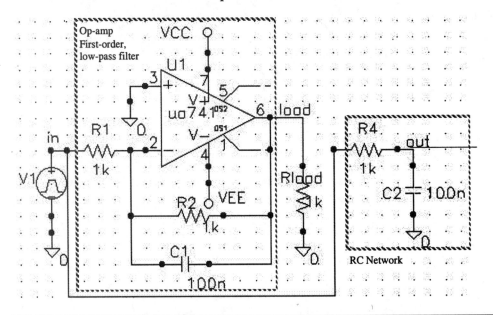

Fig. P5-1. Two first-order, low-pass filters driven by the same voltage source, V1. On the right is a simple RC circuit whose time constant is 100 μs. On the left is an op-amp enhanced version of the same circuit: an inverting amplifier with a gain of 1 buffers the output.

Tasks
Task 5.0. Preparation
  5.0.1. Review the reference material cited above for each type of low-pass filter.
  5.0.2. Create the PSPICE Schematic in Fig. P5.1 for simulation. You will need to add sources for the rail voltages, VCC and VEE, which you should set at ±10v.
  5.0.3. Simulate this circuit in the frequency domain using the AC sweep. Print out traces for :
  • the RC circuit transfer function, V(out)/V(in)
  • the op-amp filter transfer function, V(load)/V(in)
over a **decade** frequency range of
$$10 Hz < f < 1 MHz$$
on a single graph. Attach this printout in the space provided.
  Over what range of frequencies does the op-amp filter give the same frequency dependence as the RC circuit?

_____ $< f <$ _____

Insert the printout from Task 5.0.3 here:

5.0.4. Now draw a 1KΩ load resistor across the output of the RC circuit, i.e., connected between the node (out) and ground in your schematic. Run the AC Sweep analysis again.

Print out traces for :
- the RC circuit transfer function, V(out)/V(in)
- the op-amp filter transfer function, V(load)/V(in)

over a **decade** frequency range of
$$10Hz < f < 1MHz$$

on a single graph. Attach this printout in the space provided.

Which filter is more affected by the load resistor, and why?

Answer here:

Attach printout for Task 5.0.4 here:

5.0.5. Calculate the RC time-constant of the op-amp filter. Simulate its transient response to a 1 volt amplitude square-wave with a 1 volt dc offset, whose period is approximately 5x the time constant you calculated. Print out a plot showing approximately three cycles of the input and the output on the same graph and attach it in the space provided.

Attach printout for Task 5.0.5 here:

5.0.6. Create a PSPICE Schematic file for simulating the second-order low-pass filter in Fig. P5.2.

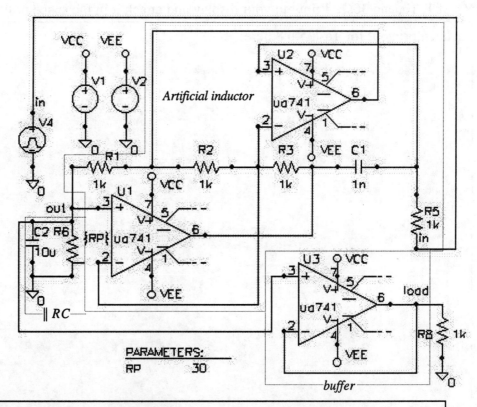

Fig. P5-2. Second-order low-pass filter circuit, employing op-amps for an artificial inductor and for buffering the load.

5.0.7. Draw below the equivalent circuit for Fig. P5-2, replacing the artificial inductor with a real inductor of the correct value. Calculate its natural frequency.

Draw equivalent circuit here:

$f_n =$ _____ Hz

5.0.8. Simulate the circuit of Fig. P5-2 with an AC Sweep from 10Hz to 1 MHz. Use a Parametric Analysis along with the AC Sweep to obtain a display with three curves for the magnitude of the filter transfer function, |V(load)/V(in)|, for values of the resistance R6 of 1, 10, and 30Ω. Print out your display and attach it in the space provided.

Attach printout for Task 5.0.8 here:

5.0.9. Identify the frequency, f_{max}, at which |V(load)/V(in)| is a maximum for the case R6 = 30 Ω, and compare it with the resonance frequency you calculated in 7, above.

5.0.10. Obtain a display of |V(load)/V(in)| vs. frequency with a dB scale for the vertical axis. [**Add Trace** db(V(load)/V(in)) to a new plot]. Use the cursors to measure the roll-off rate in dB/decade in the stop-band. This rate should be independent of R6. Enter the roll-off rate here:

Roll-off rate = _____ dB/decade.

Task 5.1. Transient and frequency response of the first-order low-pass filter.

5.1.1. Construct the first-order op-amp filter in Fig. P5.1. Use standard components with values as close as possible to those in the figure. In consultation with your lab instructor, choose a method to measure the time constant of its transient response to a step function input that starts at 0 volts and ends at 1 volt.

Time Constant Measurement Methods

Method	Reference
Virtual Bench Oscilloscope	M2.1.2
LabVIEW automated time-constant measurement	Lab Skill Exercise M2-2
LabVIEW exponential curve-fit	Lab Skill Exercise M2-3

5.1.2. Draw your measurement circuit here:

Measurement circuit:

5.1.3. Measure the time-constant. To get the data you need to obtain the time-constant of the transient response, you need **either**
- **two** *Virtual Bench* Oscilloscope screens, or
- **one** front panel screen from **either** of the VI's in Lab Skill Exercise M2-2 or M2-3.

Print out the screens necessary for your measurement method and attach them on this page and the following page.

Attach printout from Task 5.1.3 here:

Attach printout from Task 5.1.3 here (if necessary):

If you are using the *Virtual Bench* Oscilloscope, calculate the time-constant here and enter your result.

$\tau =$ _____ sec.

5.1.4. Consult your lab instructor and choose one of the methods below to measure the transfer function of your filter. Circle your choice.

Transfer function measurement methods for Task 5.1.4.

	Method	Workbook section
1	*Virtual Bench* Oscilloscope	M4.1
2	*LabVIEW* Single-frequency transfer function	M4.2.1
3	*LabVIEW* Broad-band transfer function	M4.2.2

5.1.5. Draw your measurement circuit here. Show all analog input and output channel connections:

Measurement circuit:

5.1.6. Measure the magnitude and phase of the transfer function over a sufficient frequency range to show two decades of frequency on either side of the corner frequency, $f_c = 1/$(time constant). If you are using method 1 or 2, fill in the appropriate columns in the Data Table. If you are using method 3, print out the front panel showing your results and attach it *to this page*, over the table and graph.

Data Table for Task 5.1.6

Freq. (Hz)	*Virtual Bench* Scope measurements		*Virtual Bench* calculations / *LabVIEW* measurements	
	Input Voltage Amplitude (v)	Output Voltage Amplitude (v)	|H|	Phase of H (deg)

Plot the magnitude and phase (two curves) from the Data Table on the graph below. Put the magnitude axis on the left and the phase axis on the right. Label all axes.

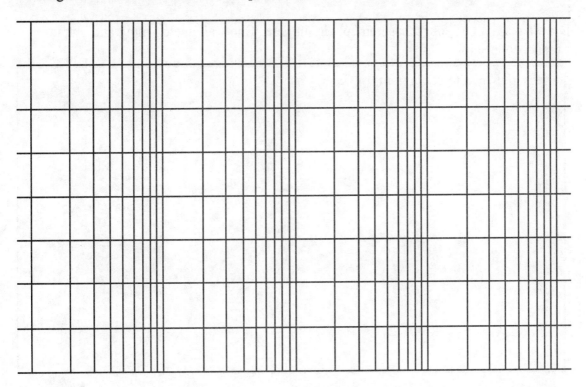

Task 5.2. Transient and frequency response of the second-order low-pass filter: under-damped case

Construct the second-order op-amp filter in Fig. P5-2. Use standard components with values as close as possible to those in the figure. Use a value of 30 Ω for the resistor R6.

5.2.1. Set-up. Consult your lab instructor and select one of the methods below for producing a low-frequency square-wave input waveform for your transient response measurement. Circle your choice.

	Method	Reference
1	*Virtual Bench* or freestanding Function Generator	M1.2
2	*LabVIEW* function generator (**FuncGen.vi**) Lab Skill Exercise M1-1	M1.3

Also choose one of the methods below for acquiring and displaying the input and output voltage waveforms. Circle your choice.

	Method	Reference
1	*Virtual Bench* or freestanding Oscilloscope	M2.1.2, M2.1.3
2	*LabVIEW* (analog software trigger) acquisition	M2.2.2

Draw a diagram of your measurement circuit here. Show all analog input and output channel connections:

Measurement circuit diagram:

5.2.2. Waveform acquisition. Use your signal acquisition instrument to acquire the input and output waveforms on the data display. The following features should appear on your waveform display:

Waveform:	Feature(s):	Present (y/n)
input	Rising edge of square wave (near left edge of screen)	
output	Decay to very close to steady state value (near right edge of screen)	

5.2.3. Cursor placement. Place cursors on your acquired waveforms as shown below:

Cursor No.	Waveform	Feature to locate:
1	output	First maximum peak after rising edge of input square wave
2	output	Second maximum peak after rising edge of input square wave
3 (*LabVIEW*)	output	Third maximum peak after rising edge of input square wave

5.2.4. Record results. Print out a copy of your instrument screen and attach it in the space provided.

Attach your instrument screen from Task 5.2.4 here:

5.2.5. Use the cursors to measure the period of the oscillations in the output. Enter the result here:

T = _____ . How does this period compare with that predicted in the simulation?

5.2.6. For oscillating waveforms like the output in this case, the "10% to 90% risetime" associated with a transition from the initial value, V_{init}, to the final value, V_{final}, is defined as the time it takes for the output, $v(t)$, to go from the level
$$V_{10\%} = V_{init} + 0.1 \times (V_{final} - V_{init})$$
to a point where it stays within the range
$$|v(t) - V_{final}| \leq 0.1 \times |V_{final} - V_{init}|$$
Set the Cursors on the appropriate points of the output waveform to measure this risetime for your filter.

Record your measured risetime here: t_R = _____ .

5.2.7. Use the cursors to measure the time constant of the exponential envelope of the oscillations. Place a pair of cursors on two successive pairs of peaks and enter the positions in the table below:

Cursor pair placement	dV	dT
1		
2		

Calculate the time constant here and enter the result:

τ = _____

5.2.8. A decaying sine wave can be expressed as an exponential of the form,
$$\exp\{(-\sigma \pm j\omega)t\}.$$
Calculate values for σ and ω for the measured output of your filter. Compare the measured ω to the calculated natural frequency from Task 5.0.7.

Calculated f_n	
Measured $\omega/(2\pi)$	

5.2.9. Consult your lab instructor and choose one of the methods below to measure the transfer function of your circuit. Circle your choice.

Transfer function measurement methods for Task 5.2.9.

	Method	Workbook section
1	*Virtual Bench* Oscilloscope	M4.1
2	*LabVIEW* Single-frequency transfer function	M4.2.1
3	*LabVIEW* Broad-band transfer function	M4.2.2

5.2.10. Draw your measurement circuit here. Show all analog input and output channel connections.

Measurement circuit:

5.2.11. While taking measurement data, you will need to make sure your frequency range covers the following features:

Transfer function part:	Feature(s)
Magnitude	From $f = 0$ up to where $|H|$ drops to 10% of its value at $f = 0$.
Phase	From $f = 0$ up to where $|H|$ drops to 10% of its value at $f = 0$.

5.2.12. Measure the magnitude and phase of the transfer function over the required frequency range. If you are using method 1 or 2, fill in the appropriate columns in the Data Table. If you are using method 3, print out the front panel showing your results and attach it to this page, over the table and graph.

Data Table for Task 5.2.12

Freq. (Hz)	Virtual Bench Scope measurements		Virtual Bench calculations / LabVIEW measurements	
	Input Voltage Amplitude (v)	Output Voltage Amplitude (v)	|H|	Phase of H (deg)

Plot the magnitude and phase (two curves) from the Data Table on the graph below. Put the magnitude axis on the left and the phase axis on the right. Label all axes.

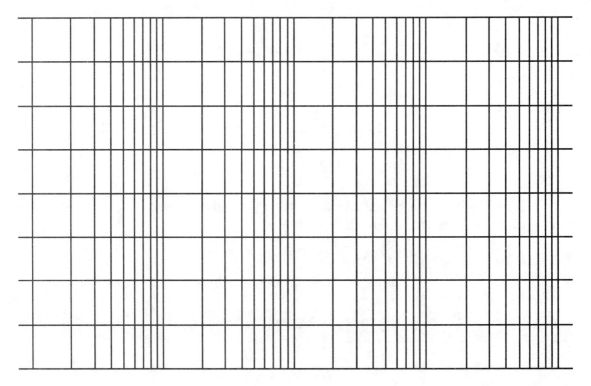

Task 5.3. Transient and frequency response of the second-order low-pass filter: Near critical damping

Modify the second-order op-amp filter in Fig. P5-2. Use a value of 10 Ω for the resistor R6.

5.3.1. Set-up. Consult your lab instructor and select one of the methods below for producing the input waveform for your transient response measurement. Circle your choice.

	Method	Reference
1	*Virtual Bench* or freestanding Function Generator	M1.2
2	*LabVIEW* function generator (**FuncGen.vi**) Lab Skill Exercise M1-1	M1.3

Also choose one of the methods below for acquiring and displaying the input and output voltage waveforms. Circle your choice.

	Method	Reference
1	*Virtual Bench* or freestanding Oscilloscope	M2.1.2, M2.1.3
2	*LabVIEW* (analog software trigger) acquisition	M2.2.2

Draw a diagram of your measurement circuit here. Show all analog input and output channel connections.

Measurement circuit diagram:

5.3.3. Waveform acquisition. Use your signal acquisition instrument to acquire the input and output waveforms on the data display. The following features should appear on your waveform display:

Waveform:	Feature(s):	Present (y/n)
input	A rising edge of the square wave but no falling edges	
output	Reaches its steady-state value	

5.3.4. Cursor placement. Place cursors on your acquired waveforms as shown below:

Cursor No.	Waveform	Feature to locate:
1	output	Where output first passes through a point 10% of the difference between its initial and its steady-state value
2	output	Where output finally stays within 10% of its steady-state value.

5.3.5. Record results. Print out a copy of your instrument screen and attach it in the space provided.

Attach your instrument screen from Task 5.3.5 here:

5.3.6. For waveforms like the output in this case, the "10% to 90% risetime" associated with a transition from the initial value, V_{init}, to the final value, V_{final}, is still defined as the time it takes for the output, $v(t)$, to go from the level
$$V_{10\%} = V_{init} + 0.1 \times (V_{final} - V_{init})$$
to a point where it stays within the range
$$|v(t) - V_{final}| \leq 0.1 \times |V_{final} - V_{init}|,$$
just as it is for the under-damped oscillations of the output in Task 5.2., above.

In Task 5.3.4-5, you set the Cursors on the appropriate points of the output waveform to measure this risetime for your filter.

Record your measured risetime here: $\tau_R = $ _____.

For which second-order low-pass filter you have tested is this risetime shorter, the under-damped or the near critically damped?

Answer:_____

5.3.7. Consult your lab instructor and choose one of the methods below to measure the transfer function of your circuit. Circle your choice.

Transfer function measurement methods for Task 5.3.7.

	Method	Workbook section
1	*Virtual Bench* Oscilloscope	M4.1
2	*LabVIEW* Single-frequency transfer function	M4.2.1
3	*LabVIEW* Broad-band transfer function	M4.2.2

Draw your measurement circuit here. Show all analog input and output channel connections.

Measurement circuit:

5.3.8. While taking measurement data, you will need to make sure your frequency range covers the following features:

Transfer function part:	Feature(s)
Magnitude	From 10 Hz to where $|H|$ drops to 10% of its value at 10 Hz
Phase	From 10 Hz to where $|H|$ drops to 10% of its value at 10 Hz

5.3.9. Measure the magnitude and phase of the transfer function over the required frequency range. If you are using method 1 or 2, fill in the appropriate columns in the Data Table. If you are using method 3, print out the front panel showing your results and attach it to this page, over the table and graph.

Data Table for Task 5.3.9

Freq. (Hz)	Virtual Bench Scope measurements		Virtual Bench calculations / LabVIEW measurements	
	Input Voltage Amplitude (v)	Output Voltage Amplitude (v)	|H|	Phase of H (deg)

Plot the magnitude and phase (two curves) from the Data Table on the graph below. Put the magnitude axis on the left and the phase axis on the right. Label all axes.

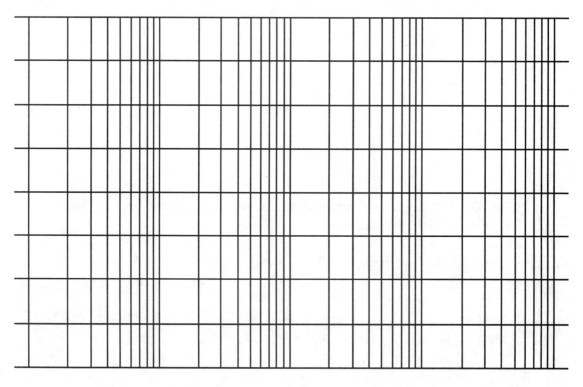

Task 5.4. Frequency response of the second-order low-pass notch filter

Modify your second-order low-pass filter circuit so that is looks like Fig. P5-3, below:

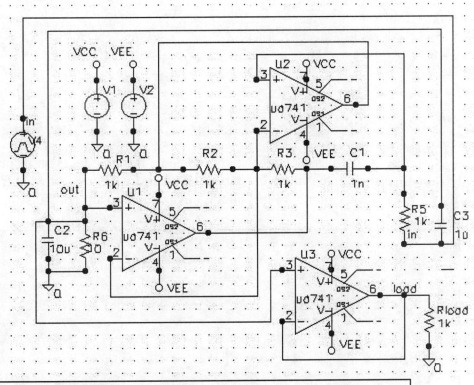

Fig. P5-3. A second order low-pass notch filter based on op-amps.

5.4.1. Draw a dotted line around the components of Fig. P5-3 which make up the artificial inductor. Draw another dotted line around the components which represent the output buffer. Label these lines.

5.4.2. Draw in the space below the equivalent circuit for Fig. P5-3, ignoring the output buffer and substituting an inductor of the appropriate value for the artificial one. Your circuit should consist of a parallel combination of an L and a C, in series with another parallel RC circuit.

5.4.3. A low-pass notch filter gets its name from the fact that its output goes to zero at a frequency in the stop band called the notch frequency. At what frequency would you expect the output of this circuit to vanish? Explain your answer.

5.4.4. Consult your lab instructor and choose one of the methods below to measure the transfer function of your circuit. Circle your choice.

Transfer function measurement methods for Task 5.4.4.

	Method	Workbook section
1	*Virtual Bench* Oscilloscope	M4.1
2	*LabVIEW* Single-frequency transfer function	M4.2.1
3	*LabVIEW* Broad-band transfer function	M4.2.2

5.4.5. Draw your measurement circuit here. Show all analog input and output channel connections.

Measurement circuit:

5.4.6. While taking measurement data, you will need to make sure your frequency range covers the following features:

Transfer function part:	Feature(s)
Magnitude	Show some of the flat region at frequencies below the "peak" and some of the flat region at frequencies above the "notch."
Phase	Same as for magnitude.

5.4.7. Measure the magnitude and phase of the transfer function over the required frequency range. If you are using method 1 or 2, fill in the appropriate columns in the Data Table. If you are using method 3, print out the front panel showing your results and attach it to this page, over the table and graph.

Data Table for Task 5.4.7

Freq. (Hz)	Virtual Bench Scope measurements		Virtual Bench calculations / LabVIEW measurements	
	Input Voltage Amplitude (v)	Output Voltage Amplitude (v)	|H|	Phase of H (deg)

Plot the magnitude and phase (two curves) from the Data Table on the graph below. Put the magnitude axis on the left and the phase axis on the right. Label all axes.

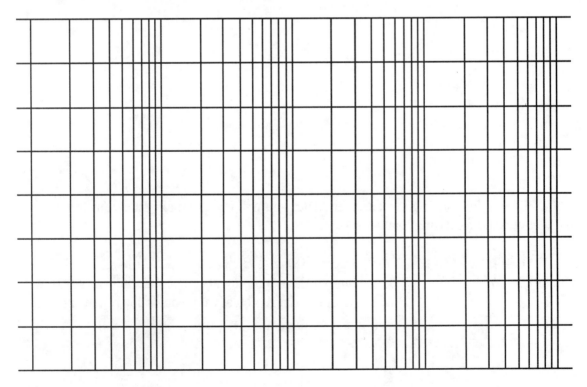

Project 6: All-pass Filters and Other Phase Shifters

Introduction

Some circuit functions require a section that acts as a pure phase shifter. These circuits have transfer functions with magnitudes near unity over the frequency range of interest, while the phase shift associated with the transfer function has a specified value, ϕ, and slope, $d\phi/df$, at some specified frequency. In this project we will consider a couple of standard all-pass filters:

Reference Table for All-pass Filters

Filter Type:	Hambly	Sedra & Smith	Nilsson & Riedel
1st order op-amp all-pass	Ch. 11 (General filter information)	Sec. 11.1, 11.4	Chs. 14 and 15 (General filter information)
2nd order op-amp all pass		Sec. 11.1, 11.4, 11.6	

First-order op-amp all-pass filter

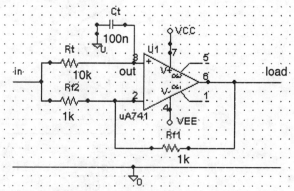

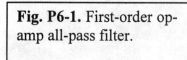

Fig. P6-1. First-order op-amp all-pass filter.

Like the low-pass filter, the all-pass has a simple first-order version which uses no op-amps at all. This filter suffers from the same shortcoming as the series RC low-pass filter, namely the magnitude of the transfer function can be strongly dependent on the load it is driving. The first-order all pass filter in Fig. P6-1 avoids this problem by having an op-amp drive the load.

The magnitude of the transfer function of the circuit in Fig. P6-1 is unity and flat at all frequencies as long as the op-amp is ideal. The phase shift is zero at low frequency, passes through –90 deg at a frequency of

$$f_0 = \frac{1}{2\pi R_t C_t}, \tag{P6-1}$$

and approaches –180 deg at high frequencies.

Second-order op-amp all-pass filter

The phase-shifter circuit in Figure P6-2 uses the artificial inductor from Project 3. A potential advantage of the more complex second-order all-pass filter over the first-order version is that the maximum available phase shift approaches 360 deg instead of 180 deg as for the first order filter.

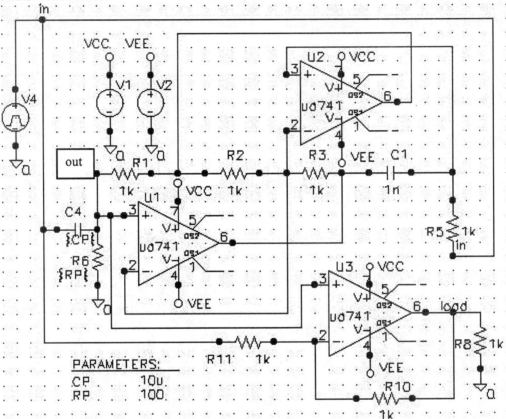

Fig. P6-2. A second-order all-pass filter, or phase shifter.

Tasks

Task 6.0. Preparation

6.0.1. Review the reference material cited above for each type of all-pass filter.

6.0.2. Create the PSPICE Schematic in Fig. P6-1 for simulation. Sources for the rail voltages, VCC and VEE, should be set at ± 10v.

6.0.3. Simulate this circuit in the frequency domain using the AC sweep. Print out traces for :
- the RC circuit transfer function, V(out)/V(in)
- the op-amp filter transfer function, V(load)/V(in)

over a **decade** frequency range of
$$10 Hz < f < 1 MHz$$
on a single graph. Attach this printout in the space provided.

Attach printout for Task 6.0.3 here:

6.0.4. Create a PSPICE Schematic file for simulating the second-order all-pass filter in Fig. P6-2. Set up the values for C4 and R6 as PARAMETERS with the nominal values shown. Set up the PARAMETRIC analysis to calculate three cases, R6 = 10, 100, and 1000 Ω. Run the AC SWEEP and PARAMETRIC analyses together, so that you get 3 curves of the phase shift p[V(load)/V(in)] versus frequency on a decade scale over the range

$$10 Hz < f < 1 MHz.$$

Print out this plot and attach it in the space provided.

All-pass filters are characterized in the frequency domain by:
• **Non-dispersive** frequency bands, in which the phase shift versus frequency is almost flat, or changes very slowly,
• **Dispersive** frequency bands, in which the phase shift changes relatively rapidly with frequency.

Mark on the plot the **dispersive** frequency bands over which the phase shifts go through a range of about $350°$ for each of the three curves. **Hint** : One of your curves is likely to look strange. Remember that phase shift is ambiguous to within $\pm 360n°$, where n is any integer. Therefore, if one of your curves looks very unlike the others, you can redraw part of it so it goes through a total phase shift of $-360°$ from low to high frequency just like the others. Redraw the strange-looking curve using this idea, and then draw in the frequency range over which the transition from ~0 to ~ -360 happens.

Does increasing the value of R6

                widen _____ or narrow _____

the **dispersive** frequency bands? (Check the right answer).

Attach printout for Task 6.0.4 here:

6.0.5. Change the PARAMETRIC analysis setup to plot 3 curves for C4 = 1, 10, and 100 µF. Run the AC SWEEP and PARAMETRIC analyses together, so that you get 3 curves of the phase shift for V(load)/V(in) versus frequency on a decade scale over the range

$$10 Hz < f < 1 MHz.$$

Note from your screen the frequencies for which the phase shift for each of the three curves passes through 180 deg, and enter those frequencies in the Data Table.

Data Table for Task 6.0.5

Value of C4	Freq. for 180 deg phase shift
1 µF	
10 µF	
100 µF	

Does increasing the value of C4

increase_____ or decrease _____

the frequency at which the phase shift crosses ~ 180 deg? (Check the right answer).

Note that the 180° frequency lies very near the middle of the **dispersive** frequency band in all three cases.

6.0.6. Disable the PARAMETRIC analysis so that R6 = 100Ω and C4 = 10 µF in this simulation. Using the TRANSIENT analysis, drive this filter with a 100 Hz square wave of 0.5 volt amplitude, with a 0.5 volt dc offset. Place the Probe Cursor on two successive peaks of the high-frequency oscillations you see in the output. Enter the times corresponding to these positions in the table below and compute the difference.

Position Number	Time reading
1	
2	
Difference = dT = oscillation period	

Calculate which harmonic of the fundamental (100 Hz) frequency these high-frequency oscillations are and enter the result here.

Harmonic number = _____.

Print out a plot showing two or three cycles of the input V(in) and the output V(load) on the same graph. Attach the plot in the space provided.

Attach your printout from Task 6.0.6 here:

6.0.7. Repeat Task 6.0.6, this time driving the filter with a 700 Hz square wave of 0.5 volt amplitude, with a 0.5 volt dc offset.

Place the Probe Cursor on two successive peaks of the high-frequency oscillations you see in the output. Enter the times corresponding to these positions in the table below, and compute the difference.

Position Number	Time reading
1	
2	
Difference = dT = oscillation period	

Calculate which harmonic of the fundamental frequency these high-frequency oscillations are, and enter the result here.

Harmonic number = _____.

6.0.8. Compare the frequencies of the oscillations you saw in Tasks 6.0.6 and 6.0.7. Do these frequencies lie within the dispersive frequency band of the all-pass filter?

Square-wave fundamental frequency	Oscillations within dispersive band (Yes/No)
100 Hz	
700 Hz	

Since this circuit is an all-pass filter and all the input frequencies pass through it without significant attenuation, you might have been tempted to assume that waveforms passing through this filter would not be significantly distorted. Qualitatively, which of the two waveforms is more distorted by this filter? (check your answer),

- _____ The 100 Hz square wave, whose higher harmonics lie within the dispersive frequency band,   OR

- _____ The 700 Hz square wave, whose lower harmonics lie within the dispersive frequency band.

Explain your answer briefly.

Task 6.1. Frequency response of the second-order all-pass filter.
Construct the second-order op-amp filter in Fig. P6-2. Use standard components with values as close as possible to those in the figure.

6.1.1. Consult your lab instructor and choose one of the methods below to measure the transfer function of your circuit. Circle your choice.

Transfer function measurement methods for Task 6.1.1.

	Method	Workbook section
1	*Virtual Bench* Oscilloscope	M4.1
2	*LabVIEW* Single-frequency transfer function	M4.2.1
3	*LabVIEW* Broad-band transfer function	M4.2.2

6.1.2. Draw your measurement circuit here. Show all analog input and output channel connections.

Measurement circuit:

6.1.3. While taking measurement data, you will need to make sure your frequency range covers the following features:

Transfer function part:	Feature(s)
Magnitude	Should be close to flat : No features expected
Phase	Include the entire dispersive band, and a little of the flat region on either side of it

6.1.4. Measure the magnitude and phase of the transfer function over the required frequency range. If you are using method 1 or 2, fill in the appropriate columns in the Data Table. If you are using method 3, print out the front panel showing your results and attach it to this page, over the table and graph.

Data Table for Task 6.1.4

Freq. (Hz)	Virtual Bench Scope measurements		Virtual Bench calculations / LabVIEW measurements	
	Input Voltage Amplitude (v)	Output Voltage Amplitude (v)	|H|	Phase of H (deg)

Plot the magnitude and phase (two curves) from the Data Table on the graph below. Put the magnitude axis on the left and the phase axis on the right. Label all axes. Identify the dispersive band.

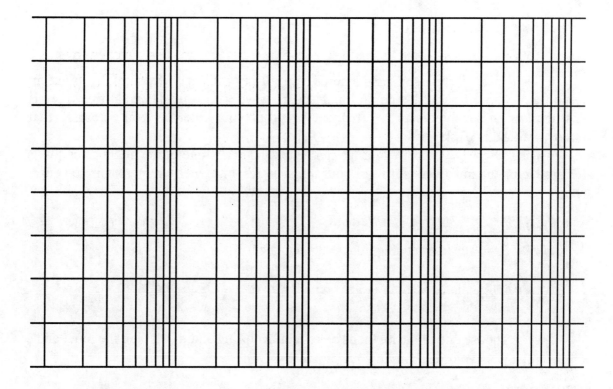

Task 6.2. Transient response of the all-pass filter. Measure the transient response of your all-pass filter to an input square-wave whose fundamental frequency is near the **low** end of the dispersive frequency band of your filter, as you measured that frequency band in Task 6.1.4.

6.2.1. Set-up. Consult your lab instructor and select one of the methods below for producing the input waveform for your transient response measurement. Circle your choice.

	Method	Reference
1	*Virtual Bench* or freestanding Function Generator	M1.2
2	*LabVIEW* function generator (**FuncGen.vi**) Lab Skill Exercise M1-1	M1.3

Also choose one of the methods below for acquiring and displaying the input and output voltage waveforms. Circle your choice.

	Method	Reference
1	*Virtual Bench* or freestanding Oscilloscope	M2.1.2, M2.1.3
2	*LabVIEW* (analog software trigger) acquisition	M2.2.2

Draw a diagram of your measurement circuit here. Show all analog input and output channel connections.

Measurement circuit diagram:

6.2.2 Waveform acquisition. Use your signal acquisition instrument to acquire the input and output waveforms on the data display. The following features should appear on your waveform display:

Waveform:	Feature(s):	Present (y/n)
input	Show 2-3 cycles of the input square wave	
output	Show 2-3 cycles of the output waveform	

6.2.3. Cursor placement. Place cursors on your acquired waveforms as shown below:

Cursor No.	Waveform	Feature to locate:
		No cursor placements required

6.2.4. Record results. Print out a copy of your instrument screen and attach it in the space provided.

Attach instrument screen from Task 6.2.4 here:

Project 7: Rectifiers

Introduction

Rectifiers are used primarily to provide DC power from an AC input. They also have applications in various wave-shaping circuits. The diode is the key building block in this type of circuit. The following circuits are but a few examples of the many that employ rectification.

Rectifier Reference Table

Rectifier type:	Hambly	Sedra and Smith
Half-wave and precision half-wave		Sec. 3.6 and 12.9
Full-wave bridge	Sec. 3.4	Sec. 3.6
Full-wave precision		Sec.12.9

Half-wave Rectifiers

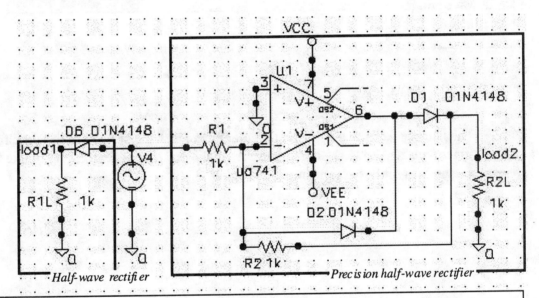

Fig. P7-1. Half-wave and precision half-wave rectifiers driven by a common sine-wave source.

Figure P7-1 shows a single sine-wave voltage source driving a simple diode half-wave rectifier to its left and a precision half-wave rectifier to its right. In both cases, what these circuits do is simply remove one half of the input sine wave (the part of the cycle where the voltage is negative) from the waveform, replacing that portion with approximately zero output.

Full-wave bridge rectifier

For producing DC power from an AC input, it is efficient to extract DC from both sides of a sine-wave input using the circuit in Fig. P7-2. The diodes provide paths for current in one direction through the load during either side of the input sine-wave cycle. The production of "clean" DC power from the output of this rectifier is essentially a low-pass filtering problem, which consists of removing "ripple" arising from the non-zero frequencies in the output.

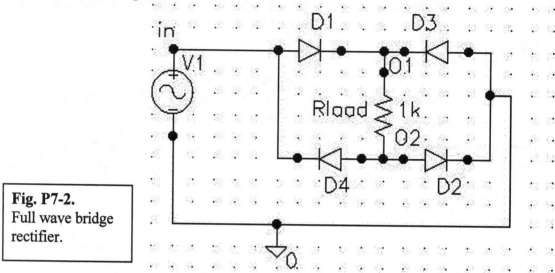

Fig. P7-2. Full wave bridge rectifier.

Full-wave precision rectifier

The non-ideal characteristics of real diodes can hurt the performance of the full-wave bridges rectifier above. Using op-amps to make real diodes perform more like ideal ones results in the rectifier design shown in Fig. P7-3.

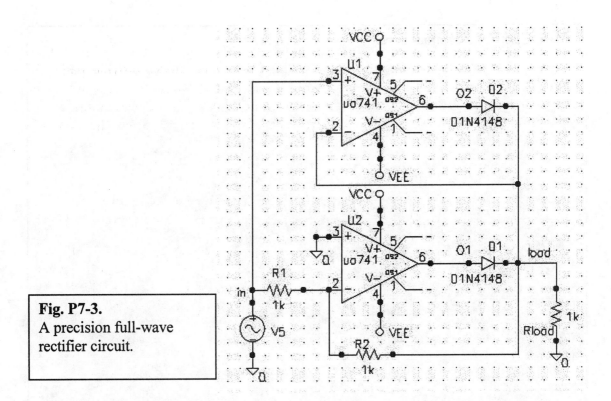

Fig. P7-3.
A precision full-wave rectifier circuit.

The op-amps in this circuit mask some of the diodes' non-ideal behavior, specifically the fact that a typical real diode needs about 0.7 volts of forward bias before it will conduct appreciable current.

Tasks
Task 7.0. Preparation
    7.0.1. Review the reference material cited above for each type of rectifier.
    7.0.2. Create the PSPICE Schematic in Fig. P7-1 for simulation. You will need to add sources for the rail voltages, VCC and VEE, which you should set at ± 10v.
    7.0.3. Set the input voltage amplitude to 1.5 volts and the sine-wave frequency to 1.0 KHz. Simulate this circuit in the time domain using the Transient Analysis. Print out a single screen containing the following:
- A plot showing three cycles of:
  - the input voltage waveform
  - the output waveform of the half-wave rectifier
  - the output waveform of the precision half wave rectifier
- A second plot showing the currents in all three diodes.

Attach this printout in the space provided.

Attach Task 7.0.3 printout here:

7.0.4. Draw in the space below the equivalent circuit of the precision half-wave rectifier when diode D1 is conducting. Explain very briefly the operation of the circuit under these conditions.

7.0.5. Draw in the space below the equivalent circuit of the precision half-wave rectifier when diode D2 is conducting. Explain very briefly the operation of the circuit under these conditions.

7.0.6. Compare the two rectifiers at low input levels. Change the input voltage amplitude to 0.5 volts and run your simulation from Task 7.0.3 again. Obtain a printout similar to Task 7.0.3 and attach it in the space provided. Which of the two rectifiers performs best at this low input voltage, and why?

Attach Task 7.0.6 printout here:

7.0.7. Compare the two rectifiers when they drive loads with high power requirements. Set the input voltage amplitude to 1.5 volts, but change both load resistances to 10Ω, and run your simulation from Task. 7.0.3 again. Print out a single screen containing the following:
- A plot showing three cycles of:
  - the input voltage
  - the output of the half-wave rectifier
  - the output of the precision half wave rectifier
- A second plot showing the total op-amp output current.

Attach it in the space provided. Which of the two rectifiers performs best at high output power, and why?

Attach Task 7.0.7 printout here:

7.0.8. Simulate two-three cycles of the input and output of the precision full-wave rectifier of Fig. P7-3. You will need to add sources for the rail voltages, VCC and VEE, which you should set at ± 10v. Use an input sine-wave of 1.5v.mplitude and 1 kHz frequency. Print out a single sheet containing
- A plot showing three cycles of:
    - the input voltage
    - the output of the rectifier
- A second plot showing the currents both diodes

Attach this graph in then space provided.

Attach Task 7.0.8 printout here:

7.0.9. Draw in the space below the equivalent circuit of the precision half-wave rectifier when diode D1 is conducting. Explain very briefly the operation of the circuit under these conditions.

7.0.10. Draw in the space below the equivalent circuit of the precision half-wave rectifier when diode D2 is conducting. Explain very briefly the operation of the circuit under these conditions.

Task 7.1. Precision Half-Wave Rectifier measurements

7.1.1. Construct the precision half-wave rectifier circuit in Fig. P7-1. Measure its transient response to a 1.5 volt amplitude, 1 KHz sine wave.

7.1.2. Set-up. Consult your lab instructor and select one of the methods below for producing the input waveform for your transient response measurement. Circle your choice.

	Method	Reference
1	*Virtual Bench* or freestanding Function Generator	M1.2
2	*LabVIEW* function generator (**FuncGen.vi**) Lab Skill Exercise M1-1	M1.3

Also choose one of the methods below for acquiring and displaying the input and output voltage waveforms. Circle your choice.

	Method	Reference
1	*Virtual Bench* or freestanding Oscilloscope	M2.1.2, M2.1.3
2	*LabVIEW* (analog software trigger) acquisition	M2.2.2

Draw a diagram of your measurement circuit here. Show all analog input and output channel connections.

Measurement circuit diagram:

7.1.3. Waveform acquisition. Use your signal acquisition instrument to acquire the input and output waveforms on the data display. The following features should appear on your waveform display:

Waveform:	Feature(s):	Present (y/n)
input	2-3 cycles	
output	2-3 cycles	

7.1.4. Cursor placement. Place cursors on your acquired waveforms as shown below:

Cursor No.	Waveform	Feature to locate:
1	output	Beginning of flat portion of output
2	output	End of flat portion of output

7.1.5. Record results. Print out a copy of your instrument screen and attach it in the space provided.

Attach your instrument screen from Task 7.1.5 here:

7.1.6. Ideally, a half-wave rectified sine wave should be flat for 50% of its period. Calculate the percent of the period that your output is flat:

Percent of cycle that is flat = _____.

7.1.7. Linear circuits generally attenuate and phase-shift differently each input frequency they see, but they do not produce output at any frequency unless driven by an input at that frequency. Non-linear circuits, such as the rectifiers in this project are different: Even when driven with a sine wave at a single frequency, they generally output frequencies at harmonics of that input frequency. Capture and display the amplitude spectrum of **exactly** four cycles of the output of your rectifier circuit, using a 1 volt amplitude, 1 KHz frequency sine wave as input. Choose one of the following methods:

	Method	Workbook Reference
1	*Virtual Bench* Digital Spectrum Analyzer	M3.2
2	*LabVIEW* Spectrum Analyzer VI	M3.3

Print out a front panel of your instrument showing the time waveform and the amplitude spectrum, and attach it in the space provided. Choose a sampling rate fast enough so you could see spectral components up to 10 KHz. Use a logarithmic or dB scale on the vertical axis of your amplitude spectrum plot. Identify on your printout all the harmonics of the 1 KHz input you can see.

Attach Task 7.1.7 printout here:

Task 7.2. Precision Full-Wave Rectifier Measurements

7.2.1. Construct the precision full-wave rectifier circuit in Fig. P7-3. Measure its transient response to a 1.5 volt amplitude, 1 KHz sine wave.

7.2.2. Set-up. Consult your lab instructor and select one of the methods below for producing the input waveform for your transient response measurement. Circle your choice.

	Method	Reference
1	*Virtual Bench* or freestanding Function Generator	M1.2
2	*LabVIEW* function generator (**FuncGen.vi**) Lab Skill Exercise M1-1	M1.3

Also choose one of the methods below for acquiring and displaying the input and output voltage waveforms. Circle your choice.

	Method	Reference
1	*Virtual Bench* or freestanding Oscilloscope	M2.1.2, M2.1.3
2	*LabVIEW* (analog software trigger) acquisition	M2.2.2

Draw a diagram of your measurement circuit here. Show all analog input and output channel connections.

Measurement circuit diagram:

7.2.3. Waveform acquisition. Use your signal acquisition instrument to acquire the input and output waveforms on the data display. The following features should appear on your waveform display:

Waveform:	Feature(s):	Present (y/n)
input	2-3 cycles	
output	2-3 cycles	

7.2.4. Cursor placement. Place cursors on your acquired waveforms as shown below:

Cursor No.	Waveform	Feature to locate:
	No cursors required	

7.2.5. Record results. Print out a copy of your instrument screen and attach it in the space provided.

Attach your instrument screen from Task 7.2.5 here:

7.2.6. Linear circuits generally attenuate and phase-shift differently each input frequency they see, but they do not produce output at any frequency unless driven by an input at that frequency. Non-linear circuits, such as the rectifiers in this project, are different: even when driven with a sine wave at a single frequency, they generally output frequencies at harmonics of that input frequency. Capture and display the amplitude spectrum of **exactly** four cycles of the output of your precision full-wave rectifier circuit, using a 1 volt amplitude, 1 KHz frequency sine wave as input. Choose one of the following methods:

	Method	Workbook Reference
1	*Virtual Bench* Digital Spectrum Analyzer	M3.2
2	*LabVIEW* Spectrum Analyzer VI	M3.3

Print out a front panel of your instrument showing the time waveform and the amplitude spectrum and attach it in the space provided. Choose a sampling rate fast enough so you could see spectral components up to 10 KHz. Use a logarithmic or dB scale on the vertical axis of your amplitude spectrum plot. Identify on your printout all the harmonics of the 1 KHz input you can see.

Attach Task 7.2.6 printout here:

Project 8: Limiting Circuits

Introduction

The purpose of a limiter is explained pretty well by just its name. In the laboratory, you may want to protect delicate measuring circuitry from an unpredictable input signal. Or, you may want to prevent the amplitude of some oscillations in a circuit from growing too large. In general, the output voltage, v_o, of a limiter will depend on the input voltage, v_i, as shown in Fig. P8-1. The output will be linear within some defined range (although the constant of proportionality K may take on any value, including negative ones) and will depend relatively weakly on v_i outside that range. The limiter is called a "hard" limiter if the transition at the boundaries of the defined range is abrupt and a "soft" limiter if the transition is gradual. These terms are qualitative and inexact.

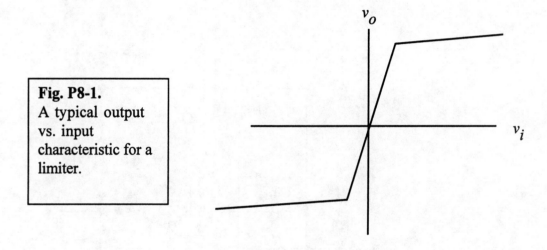

Fig. P8-1. A typical output vs. input characteristic for a limiter.

Limiter Reference Table:

Limiter Type:	Hambly	Sedra & Smith
Passive limiter	Section 3.5	Section 3.7
Active limiter		Section 12.1

Passive limiters

These simple circuits make use of the transition a diode makes from near-open to near-short as the voltage across it passes through the "on-voltage" of approximately 0.7 v. Other limiters utilize the breakdown region of a Zener diode, where voltage remains near a designed value over a large range of current. The operation of these circuits is straightforward. As a more interesting laboratory example, consider the active limiter employing an op-amp.

Active Limiters

Since an op-amp's output voltage is limited by its supply voltages (the "rails," V_{CC} and V_{EE}), any amplifier which uses an op-amp is a form of active limiter. The active limiter shown in Fig. P8-2 is very flexible and useful. Clearly, when both diodes are open, the circuit behaves like an inverting amplifier with a gain of
$-R_f / R_1$.

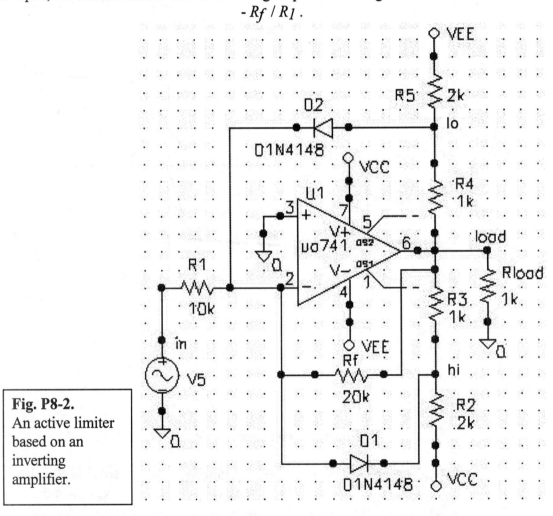

Fig. P8-2. An active limiter based on an inverting amplifier.

When either diode turns "on," it places an additional resistor (R_3 or R_4) in parallel with the feedback resistor R_f. This decreases the feedback resistance in the inverting amplifier and therefore decreases its gain in the regions outside the defined range of the limiter.

Tasks
Task 8.0. Preparation
    8.0.1. Review the reference material on limiting circuits cited above.
    8.0.2. Create the schematic Fig. P8-2 in PSPICE. Include ± 10v dc sources for VEE and VCC. Perform a DC Sweep of the input voltage from -9.5 to +9.5 v. Print out a single plot showing:
- the voltage at the load (one Y-axis)
- the currents in the two diodes (second Y-axis)

Attach this plot in the space provided. Label on your plot with the numbers 1, 2, and 3, the three ranges of input voltage that correspond to different combinations of diode states.

Attach Task 8.0.2 printout here:

8.0.3. Draw in the space below the equivalent circuits corresponding to the three ranges of input voltages you labeled in Task 8.0.2. State in the Table whether the diodes D1 and D2 are on or off in each region:

• Equivalent circuit for Region 1:

Reg.	D1	D2
1		
2		
3		

• Equivalent circuit for region 2:

• Equivalent circuit for region 3:

8.0.4. Suppose the input voltage is exactly at the low end of the central region of your plot from Task 8.0.2. If the diode D2 that is just turning off or on at this voltage were ideal, the voltage across it would be zero for this input. Hence, it would put the node **lo** in Fig P8-2 at zero potential, since the inverting op-amp input is a virtual ground. Find an expression in terms of V_{EE}, R4 and R5, for the value the output voltage V(load) must have in order to put the node **lo** at zero volts. This value is called the negative limiting level for the limiter. Put your calculation here:

8.0.5. Suppose the input voltage is exactly at the high end of the central region of your plot from Task 8.0.2. If the diode D1 that is just turning off or on at this voltage were ideal, the voltage across it would be zero for this input. Hence it would put the node **hi** in Fig P8-2 at zero potential, since the inverting op-amp input is a virtual ground. Find an expression in terms of V_{CC}, R2 and R3, for the value the output voltage V(load) must have in order to put the node **hi** at zero volts. This value is called the positive limiting level for the limiter. Put your calculation here:

8.0.6. Simulate in PSPICE the output of the limiter in Fig. P8-2 when the input is an exponentially growing sine wave. [See the VSIN voltage source and its attributes in the Appendix.] Perform a Transient analysis of the output waveform and observe how the output is limited, or clipped, at the positive and negative limiting levels.

8.0.7. Simulate the use of a limiter to shape a triangle-wave into something resembling a sine-wave. Instead of the sine-wave source V5 in Fig. P8-2, substitute a 1KHz triangle wave of 5v amplitude. [There are no triangle wave voltage sources in PSPICE. You will have to make do with a piecewise linear (VPWL) or pulse (VPULSE) source with the attributes set so that the waveform looks like a triangle wave. See these sources in the Appendix.] Also, put in new values of R2 and R5 so as to change the negative and positive limiting levels to $\pm 2v$. Perform a Transient analysis. Print out a plot showing about three cycles of the input and output and attach it in the space provided.

Attach Task 8.0.7 printout here:

8.0.8. Examine the Fourier spectra of the triangle wave, V(in) and the more sinusoidal V(load) from Task 8.0.7. Adjust the X-axis of the plot so that it covers the frequency range, $0 < f < 20\text{KHz}$. Print out a screen with two plots:
- the Fourier spectrum of V(in)
- the Fourier spectrum of V(load)

Note that the approximate sine-wave you constructed with the limiter still contains higher harmonic frequencies, although their amplitudes are substantially lower than the corresponding frequencies in the spectrum of the triangle wave. Attach your printout in the space provided.

AttachTask 8.0.8 printout here:

Task 8.1. Limiter circuit with sine wave input

8.1.1. Construct a limiter circuit with the design in Fig. P8-2. Use standard resistors as close as possible to the following values:

R1 = Rf = 10KΩ
R3 = R4 = 1KΩ
R2 = R5 = 5KΩ

Set the rail voltages to $\pm$10v. Supply a 5 KHz sine-wave input.

8.1.2. Set-up. Consult your lab instructor and select one of the methods below for producing the input waveform for your transient response measurement. Circle your choice.

	Method	Reference
1	*Virtual Bench* or freestanding Function Generator	M1.2
2	*LabVIEW* function generator (**FuncGen.vi**) Lab Skill Exercise M1-1	M1.3

Also choose one of the methods below for acquiring and displaying the input and output voltage waveforms. Circle your choice.

	Method	Reference
1	*Virtual Bench* or freestanding Oscilloscope	M2.1.2, M2.1.3
2	*LabVIEW* (analog software trigger) acquisition	M2.2.2

Draw a diagram of your measurement circuit here. Show all analog input and output channel connections:

Measurement circuit diagram:

8.1.3. Waveform acquisition. Use your signal acquisition instrument to acquire the input and output waveforms on the data display. The following features should appear on your waveform display:

Waveform:	Feature(s):	Present (y/n)
input	2-3 cycles. Increase amplitude of the input until you can just begin to see visible limiting of the output	
output	2-3 cycles	

8.1.4. Cursor placement. Place cursors on your acquired waveforms as shown below:

Cursor No.	Waveform	Feature to locate:
1	output	Positive limiting level
2	output	Negative limiting level

8.1.5. Record results. Print out a copy of your instrument screen and attach it in the space provided.

Attach your instrument screen from Task 8.1.5 here:

8.1.6. Linear circuits generally attenuate and phase-shift differently each input frequency they see, but do not produce output at any frequency unless driven by an input at that frequency. Non-linear circuits, such as the limiter in this project are different: even when driven with a sine wave at a single frequency, they generally output frequencies at harmonics of that input frequency. Capture and display the amplitude spectrum of **exactly** four cycles of the output of your rectifier circuit, using a 1 volt amplitude, 1 KHz frequency sine wave as input. Choose one of the following methods:

	Method	Workbook Reference
1	*Virtual Bench* Digital Spectrum Analyzer	M3.2
2	*LabVIEW* Spectrum Analyzer VI	M3.3

Print out a front panel of your instrument showing the time waveform and the amplitude spectrum and attach it in the space provided. Choose a sampling rate fast enough so you could see spectral components up to 10 KHz. Use a logarithmic or dB scale on the vertical axis of your amplitude spectrum plot. Identify on your printout all the harmonics of the 1 KHz input you can see.

Attach Task 8.1.6 printout here:

Task 8.2. Limiter circuit with triangle input.

8.2.1. Provide a 2 KHz *triangle* wave input to the limiter. Consult your lab instructor and select one of the methods below for producing the input waveform for output spectrum measurement. Circle your choice.

	Method	Reference
1	*Virtual Bench* or freestanding Function Generator	M1.2
2	*LabVIEW* function generator (**FuncGen.vi**) Lab Skill Exercise M1-1	M1.3

8.2.2. Capture and display the amplitude spectrum of **exactly** four cycles of the output of your circuit. Choose one of the following methods:

	Method	Workbook Reference
1	*Virtual Bench* Digital Spectrum Analyzer	M3.2
2	*LabVIEW* Spectrum Analyzer VI	M3.3

Adjust the amplitude of the input to **minimize** the higher harmonics generated by the nonlinearity of the limiter. Choose a sampling rate fast enough so you could see spectral components up to 20 KHz. Use a logarithmic or dB scale on the vertical axis of your amplitude spectrum plot. Print out a front panel of your instrument showing the time waveform and the amplitude spectrum and attach it in the space provided. Identify on your printout all the harmonics of the 2 KHz fundamental that you can see.

This task is an example of using a limiter to shape a waveform. In this case, you modified a triangle wave input to make it look more like a sine wave.

Attach Task 8.2.2 printout here:

Project 9: RC Relaxation Oscillator

Introduction

One way of making an oscillator would be to cause an op-amp to drive its output alternately between the two supply voltages, V_{CC} and V_{EE}. The circuit in Fig. P9-1 does this by feeding back two signals to the op-amp inputs. First consider the signal fed back to the positive input to the op-amp. The resistors R3 and R2 form a voltage

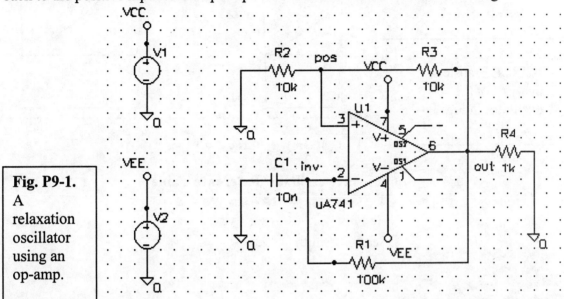

Fig. P9-1. A relaxation oscillator using an op-amp.

divider that provides exactly half the output, V(out), at the positive (non-inverting) input node, **pos** . Thus, V(pos) = V(out)/2 at all times. Next, consider the signal fed back to the inverting input node, **inv**. That feedback network consists of R1 in series with C1. The voltage across C1 (which also appears at the inverting input of the op-amp) can charge all the way up to V(out) *but will not do so immediately.* Rather, it will move exponentially toward V(out) with a time constant given by the product of R1 and C1. The voltage at node, **inv**, eventually catches up with and then surpasses that at node, **pos**. When that happens, the difference between V(pos) and V(inv) changes sign, and the op-amp must then swing its output to the opposite supply. There, the process starts all over again. Figure P9-2 shows a PSPICE simulation of the circuit in Fig. P9-1 in the time-domain. You can see how, every time the input voltage difference V(pos) - V(inv) changes sign, the op-amp output goes to the opposite rail.

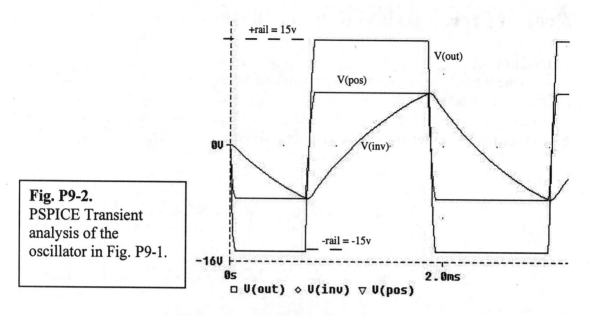

Fig. P9-2.
PSPICE Transient analysis of the oscillator in Fig. P9-1.

Square-wave oscillator Reference Table

Text	Section
Hambly	12.2
Sedra & Smith	12.5

Tasks

Task 9.0. Preparation

9.0.1. Review the reference material on square-wave oscillators.

9.0.2. Simulate the circuit of Fig. P9-1 in the time-domain, using a Transient analysis in PSPICE. Set VCC=10v and VEE=-10v. Obtain a single plot showing two or three cycles of V(out), V(pos), and V(inv).

Now, slow down this simulated oscillator by increasing the value of the capacitor C1 to 50 nF. Print out a single plot showing two or three cycles of V(out), V(pos), and V(in). Attach your plot in the space provided.

Attach Task 9.0.2 printout here:

Task 9.1. Measuring the output waveform.

9.1.1. Setup. Choose one of the methods below for acquiring and displaying the output voltage waveform. Circle your choice.

	Method	Reference
1	*Virtual Bench* or freestanding Oscilloscope	M2.1.2, M2.1.3
2	*LabVIEW* (analog software trigger) acquisition	M2.2.2

Draw a diagram of your measurement circuit here. Show all analog input and output channel connections.

Measurement circuit diagram:

9.1.2. Waveform acquisition. Use your signal acquisition instrument to acquire the output waveforms on the data display. The following features should appear on your waveform display:

Waveform:	Feature(s):	Present (y/n)
output	2-3 cycles	
node **inv**	2-3 cycles	

9.1.3. Cursor placement. Place cursors on your acquired waveforms as shown below:

Cursor No.	Waveform	Feature to locate:
1	output	Beginning of rising edge
2	output	End of rising edge

9.1.4. Record results. Print out a copy of your instrument screen and attach it in the space provided.

Attach your instrument screen from Task 9.1.4 here:

9.1.5. The cursor positions indicate that your square wave output is not ideal: that is, the rising and falling edges are not vertical. The maximum rate at which an op-amp can change its output voltage is limited by its slew rate, the maximum rate of change in output voltage,

$$\frac{dv_{out}}{dt}$$

it can sustain. Use the cursors on your plot to calculate the slew-rate for your op-amp in this oscillator, and enter the result.

Slew rate = _____ volt/s.

9.1.6. In your PSPICE simulation of this oscillator, you saw how its frequency could be controlled by adjusting the time constant of the RC feedback network. There is another way to adjust this oscillation frequency: adjusting the division ratio in the voltage divider feedback network. Try some different values for R2 in your oscillator and measure the resulting square-wave output frequencies. Do this by acquiring the output waveform as you did in Task 9.1.1, then placing two cursors on two successive rising edges to measure the period. Enter your results in the Data Table below.

Data Table for Task 9.1.6

R2 value (Ω)	Period (s)	Frequency (Hz)

Task 9.2. Oscillator square-wave output spectrum measurement. An ideal square wave of frequency f has harmonics at higher frequencies whose amplitudes are related to each other as shown in the table below.

Frequency	Relative amplitude
f	1
$2f$	0
$3f$	1/3
$4f$	0
$5f$	1/5
$6f$	0
$7f$	1/7
$8f$	0

9.2.1. Capture and display the amplitude spectrum of **exactly** four cycles of the output of your oscillator circuit. Choose one of the following methods:

	Method	Workbook Reference
1	*Virtual Bench* Digital Spectrum Analyzer	M3.2
2	*LabVIEW* Spectrum Analyzer VI	M3.3

Print out a front panel of your instrument showing the time waveform and the amplitude spectrum, and attach it in the space provided. Choose a sampling rate fast enough so you could see spectral components up to 10 KHz. Use a logarithmic or dB scale on the vertical axis of your amplitude spectrum plot. Identify on your printout all the harmonics of the 1 KHz input you can see.

9.2.2. Use a cursor to identify the amplitudes of all the harmonic frequencies in the output spectrum you have measured. Complete the Data Table below for your oscillator.

Data Table for Task 9.2.2.

Freq.	Measured freq. (Hz)	Ideal relative amplitude	Measured amplitude	Ratio to f ampl.
f		1		1
$2f$		0		
$3f$		0.333		
$4f$		0		
$5f$		0.2		
$6f$		0		

Attach Task 9.2.1 printout here:

Task 9.3. Time-constant of the V(inv) waveform. Measure the time constant of the repeating exponential waveform V(inv) in Fig. P9-1. Consult with your lab instructor to choose one of the following methods:

Method	Reference
Virtual Bench Oscilloscope	Section M2.1.3.B
Waveform Acquisition on demand	Prep Exercise M2-6
Automated time-constant calculation	Lab Skill Exercise M2-2
Automated averaging	Lab Skill Exercise M2-3

Circle your choice.

If you used the *Virtual Bench* Oscilloscope or the Waveform Acquisition on demand, tabulate your cursor location *differences* here and show your calculations below:

Scope View No.	dV(volts)	dT(s)
1		
2		

If you used one of the other two methods, print out the front panel showing your results and attach it to this page.

Project 10: Phase-shift Oscillator

Introduction

In this project, you will build and perform measurements on another type of op-amp based oscillator which outputs a more nearly sinusoidal waveform than the square-wave oscillator studied in Project 9.

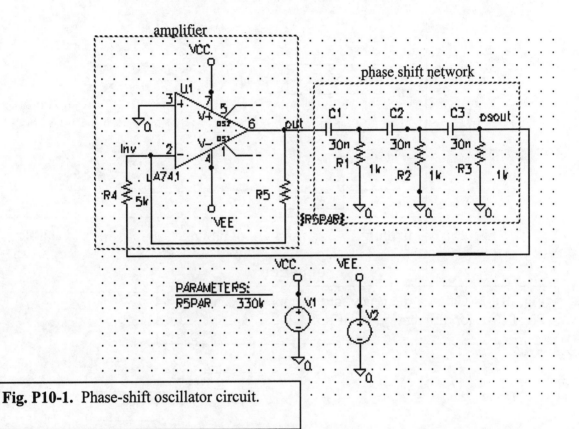

Fig. P10-1. Phase-shift oscillator circuit.

Figure P10-1 shows the circuit for the phase-shift oscillator. Its operation is best understood by dividing it into two functional blocks, the amplifier and the phase-shift network, each enclosed in a dashed box in Fig. P10-1.

The amplifier is a standard inverting amplifier, which provides a low-frequency gain given by

$$G_0 = -\frac{R5}{R4} \quad \text{(P10-1)}$$

and a phase shift of 180 deg. The phase shift network provides an additional 180 deg phase shift at the frequency of oscillation. Thus, for a signal making a round trip around the feedback loop, the total phase shift is 360 deg, or effectively zero at the oscillation frequency. The amplifier must provide sufficient gain at this oscillation frequency to overcome the loss in the phase shift network plus some additional gain to make the oscillations grow.

The gain of the amplifier section is set by the resistors, R4 and R5. With a non-ideal op-amp, gain decreases with frequency owing to the finite gain-bandwidth product of the op-amp. An example of this effect in the 741 op-amp is simulated in PSPICE and shown in Fig. P10-2, where the 330 KΩ resistor, R5, is reduced to 110 K and then raised to 550K.

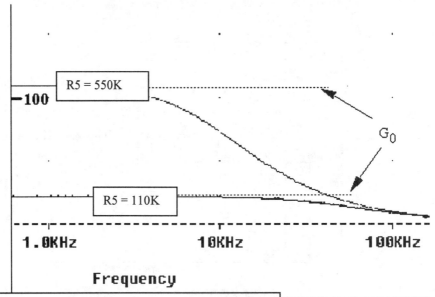

Fig. P10-2. Gain vs. frequency for the amplifier section of Fig. P10-1, for different values of R5.

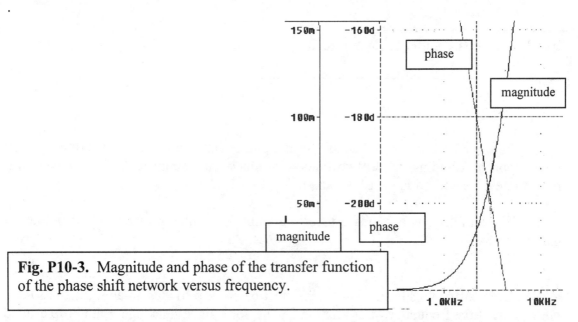

Fig. P10-3. Magnitude and phase of the transfer function of the phase shift network versus frequency.

Now, consider the phase shift network. According to the above discussion, it must provide a total phase shift of 180 deg at the oscillation frequency between the nodes **out** and **psout** in Fig. P10-1 and not attenuate the magnitude so much that you can't compensate for it with amplifier gain. Figure P10-3 shows a PSPICE plot of the magnitude and phase of the transfer function, V(psout)/V(out), of the phase shift network versus frequency. You can obtain a plot like this by connecting any PSPICE voltage source to the input (the **out** node in Fig P10-1) and performing an AC sweep analysis. The PROBE cursor in Fig. P10-3 is placed where the phase shift of the phase shift network passes through 180 deg. The plot shows that the magnitude of the signal to be fed back into the amplifier has been reduced by a factor of approximately 0.04. Therefore, the amplifier will need to have a gain of at least $(0.04)^{-1}$, or 25, for the circuit to have any chance to sustain oscillation at the oscillation frequency, which the plot indicates is approximately 2 KHz.

More detailed discussion of the phase-shift oscillator can be found in these references:

Reference Table for Phase-shift Oscillator

Reference	Section
Hambly	9.11, 9.12
Sedra & Smith	12.2

Tasks
Task 10.0. Preparation
10.0.1. Review the reference material on the phase shift oscillator.
10.0.2 PSPICE simulation of the phase shift network
Perform an AC sweep simulation of the phase shift network by itself, driven with a simple voltage source as shown in Fig. P10-4..

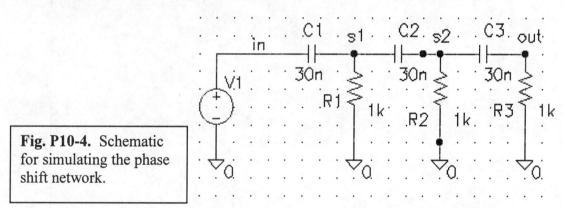

Fig. P10-4. Schematic for simulating the phase shift network.

Use an AC Sweep analysis and a decade scale of frequency. Print out a plot showing the amplitude (one y-axis) and the phase (this will take another y-axis) of V(out)/V(in) over a range of frequencies at least a decade on either side of the point where the phase shift passes through 180 deg. Use the PROBE cursor to locate the frequency where the phase is 180 deg and show the cursor on your printout. Use the PROBE Cursor to determine the magnitude of the transfer function at this frequency, and enter it here:
$|H|$ = _____

Attach your printout in the space provided.

Attach Task 10.0.2 printout here:

The frequency you measured in this task is a rough prediction of the output frequency of your oscillator. The transfer function magnitude above represents a loss through the phase shift network, which must be compensated (and then some) by the amplifier circuit.

10.0.3. Perform an AC Sweep simulation of the amplifier circuit by itself, driven with a simple voltage source as shown in Fig. P10-5.

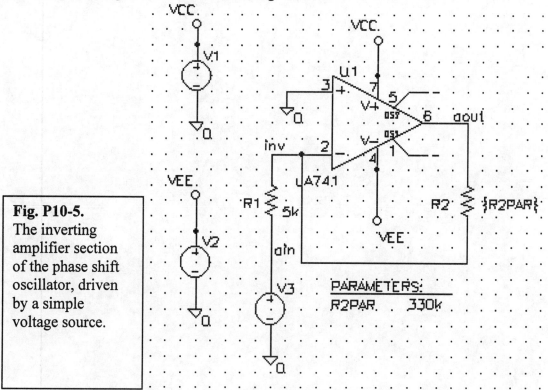

Fig. P10-5. The inverting amplifier section of the phase shift oscillator, driven by a simple voltage source.

Set VCC = +8 V and VEE = -8 V. Use a Parametric Analysis along with your AC Sweep, and let R2 take on the following list of values:
  R2 = 80k, 110k, 220k, 330k, and 500k.

Print out a family of curves which show the amplifier gain, V(aout)/V(ain) over the frequency range, $100 < f < 100$ KHz, on a decade scale. Attach the printout in the space provided.

From your results, circle which values for R2 on the list below provide enough gain in your amplifier to overcome the loss in the phase shift network and give the oscillator a chance to operate:

  R2 = 80k, 110k, 220k, 330k, and 500k.

Attach Task 10.0.3 printout here:

10.0.4. Perform a Transient Analysis of the complete oscillator circuit (Fig. 2.13). Use a Parametric Analysis along with your Transient Analysis and obtain a family of curves of V(out) vs. time for the R2 values 110K, 330K, and 500 K. Show enough time on your horizontal axis so you can see the oscillations grow to their final magnitude for the 330K and 500K values of R2. **Remember, you may need to put some initial voltage on one of the capacitors in the circuit [for example, IC=0.1 on the Attribute List of one of the capacitors]** in order to see the oscillations grow. Print out this family of curves and attach it in the space provided.

Attach Task 10.0.4 printout here:

P10-8

Task 10.1. Build and test the phase shift network.

    10.1.1. Build the phase-shift network section of the phase shift oscillator. Use standard component values as near as possible to those in Fig. P10-1. Then measure its transfer function.

    10.1.2. Consult your lab instructor and choose one of the methods below to measure the transfer function of your circuit. Circle your choice.

Transfer function measurement methods for Task 10.1.2.

	Method	Workbook section
1	*Virtual Bench* Oscilloscope	M4.1
2	*LabVIEW* Single-frequency transfer function	M4.2.1
3	*LabVIEW* Broad-band transfer function	M4.2.2

    10.1.3. Draw your measurement circuit here. Show all analog input and output channel connections.

Measurement circuit:

    10.1.4. While taking measurement data, you will need to make sure your frequency range covers the following features:

Transfer function part:	Feature(s)
Phase	Show at least one decade of frequency on either side of the frequency for which the phase shift = 180 deg

10.1.5. Measure the magnitude and phase of the transfer function over the required frequency range. If you are using method 1 or 2, fill in the appropriate columns in the Data Table. If you are using method 3, print out the front panel showing your results and attach it to this page, over the table and graph.

Data Table for Task 10.1.5

Freq. (Hz)	Virtual Bench Scope measurements		Virtual Bench calculations / LabVIEW measurements	
	Input Voltage Amplitude (v)	Output Voltage Amplitude (v)	|H|	Phase of H (deg)

Plot the magnitude and phase (two curves) from the Data Table on the graph below. Put the magnitude axis on the left and the phase axis on the right. Label all axes.

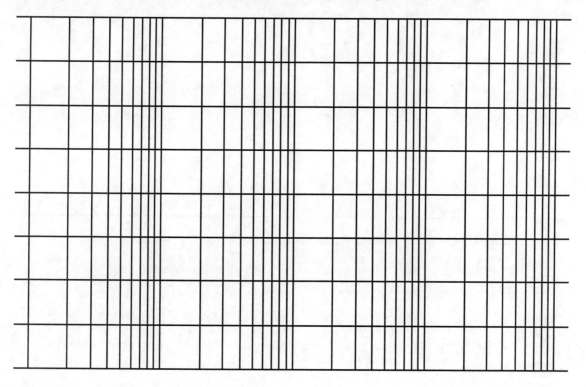

Task 10.2. Build and test the amplifier section of the phase shift oscillator. Choose a value for R2 that you think will give the amplifier enough gain so the oscillator will operate. Set VCC = +8 V and VEE = -8 V, using the DC power supplies on your lab bench. Use a sine wave input. Measure the transfer function of the amplifier over the same frequency range you measured in Task 10.1

10.2.1 Consult your lab instructor and choose one of the methods below to measure the transfer function of your circuit. Circle your choice.

Transfer function measurement methods for Task 10.2.1.

	Method	Workbook section
1	*Virtual Bench* Oscilloscope	M4.1
2	*LabVIEW* Single-frequency transfer function	M4.2.1
3	*LabVIEW* Broad-band transfer function	M4.2.2

10.2.2. Draw your measurement circuit here. Show all analog input and output channel connections:

Measurement circuit:

10.2.3. While taking measurement data, you will need to make sure your frequency range covers the following features:

Transfer function part:	Feature(s)
Magnitude	Same frequency range as the measurements on the phase shift network.
Phase	

10.2.4. Measure the magnitude and phase of the transfer function over the required frequency range. If you are using method 1 or 2, fill in the appropriate columns in the Data Table. If you are using method 3, print out the front panel showing your results and attach it to this page, over the table and graph.

Data Table for Task 10.2.4

Freq. (Hz)	Virtual Bench Scope measurements		Virtual Bench calculations / LabVIEW measurements	
	Input Voltage Amplitude (v)	Output Voltage Amplitude (v)	|H|	Phase of H (deg)

Plot the magnitude and phase (two curves) from the Data Table on the graph below. Put the magnitude axis on the left and the phase axis on the right. Label all axes.

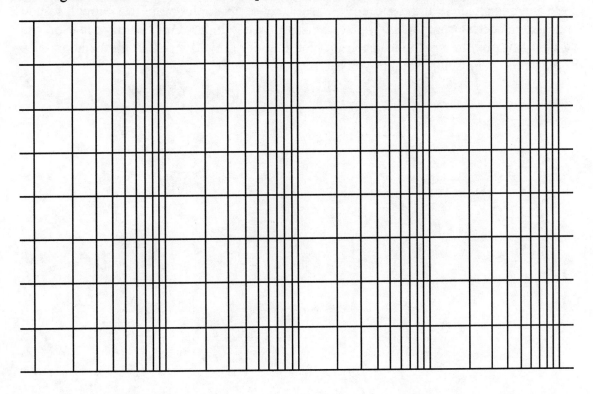

Does the amplifier gain you measured look sufficiently large to overcome the loss in the phase shift network that you measured earlier? **If it does not, go back and repeat this task with a different R2 that will provide more gain to the amplifier. Attach your new results over the old ones.**

Task 10.3. Measure the output of the oscillator in the time and frequency domains.

Capture and display the amplitude spectrum of **exactly** four cycles of the output of your oscillator. Choose one of the following methods:

	Method	Workbook Reference
1	*Virtual Bench* Digital Spectrum Analyzer	M3.2
2	*LabVIEW* Spectrum Analyzer VI	M3.3

Print out a front panel of your instrument showing the time waveform and the amplitude spectrum and attach it in the space provided. Choose a sampling rate fast enough so you could see spectral components up to 20 KHz. Use a logarithmic or dB scale on the vertical axis of your amplitude spectrum plot. Identify on your printout all the harmonics of the "almost sinusoidal" output you can see.

Attach Task 10.3 printout here:

Project 11: Amplifier circuits with Bipolar Junction Transistors (BJT's)

Introduction

Single-stage amplifiers employing BJT's have been a staple in undergraduate electrical engineering courses for many years. Many textbooks contain examples of good designs for these amplifiers. This laboratory project is based on the "universal BJT amplifier" circuit in Section 4.11 of Sedra and Smith. This circuit is slightly more complicated (i.e., it contains a few more components than may be absolutely necessary in some applications and uses both positive and negative power supply voltages) than some other common examples, but it is a useful jumping-off point for computer-aided design using PSPICE.

By making the appropriate connections to the following circuits:
- a signal source with its associated source impedance serving as the amplifier *input*,
- a load circuit (in this case just a load resistor) to be driven by the amplifier *output*, and
- a ground point,

the universal amplifier circuit in Fig. P11-1 can be operated as a common emitter, a common collector (usually called an "emitter follower"), or a common base amplifier.

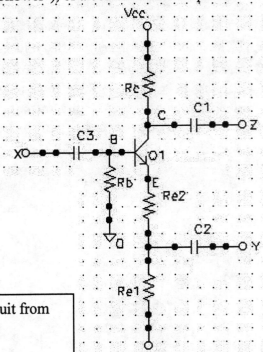

Fig. P11-1. Universal BJT amplifier circuit from Sedra and Smith.

The positive and negative power supply voltages, Vcc and Vee respectively, are always connected to the terminals indicated in Fig. P11-1. The signal source, load, and ground connection circuits look like Fig. P11-2.

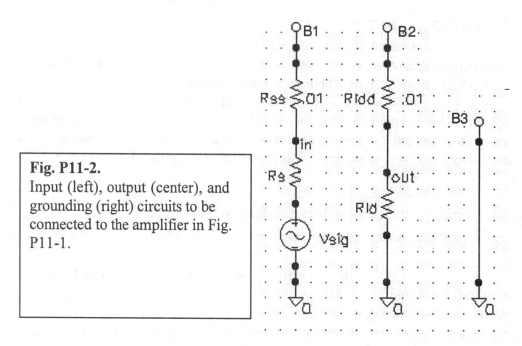

Fig. P11-2.
Input (left), output (center), and grounding (right) circuits to be connected to the amplifier in Fig. P11-1.

The input circuit consists of a signal source in series with its output impedance, Rs. The 0.01Ω resistor Rss is an artifact that allows giving two names in the PSPICE simulation to the same node. The output circuit consists of a load resistor Rld. The 0.01 Ω resistor Rldd serves the same purpose as Rss. Figure P11-2 contains **bubbles**, which you **Get** in PSPICE like any other **Schematic** part. The bubbles, B1, B2, and B3, allow you to connect these circuit points to any of the points bubbles X, Y, or Z, on the universal amplifier circuit by renaming them X, Y, or Z. Thus, different connections of the bubbles produce different amplifier circuits:

- B1->X, B2->Z, and B3->Y produce a common-emitter amplifier, and
- B1->X, B2->Y, B3->Z produce an emitter follower.

A designer usually calls on common emitter and emitter follower amplifiers to do different jobs: the common emitter amplifier can provide significant *voltage gain* when both the signal voltage and the source impedance Rs are small. However it has a larger output impedance than you might desire when you want it to drive a low value of load resistance, Rld.

The emitter follower (common-collector) amplifier does not provide voltage gain, but it can take input from a signal source with relative high source impedance, Rs, and provide sufficient *current gain* to drive a relatively low value of load resistance, Rld.

Tasks
Task 11.0. Preparation
11.0.1. Review the reference material on BJT amplifiers cited above.

11.0.2. Create a PSPICE Schematic of the universal BJT amplifier in the common-emitter configuration as shown in Fig. P11-3.

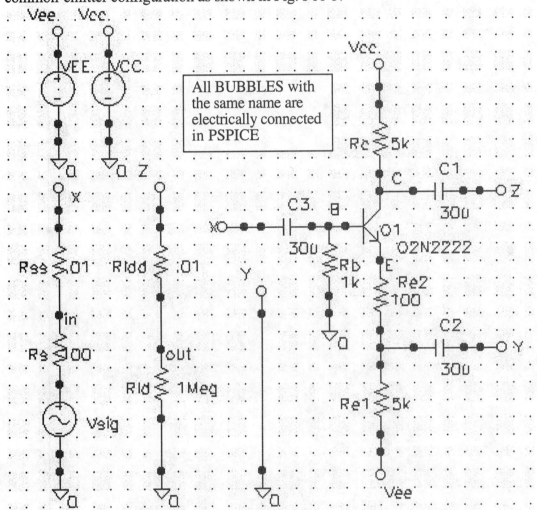

Fig. P11-3. Universal BJT amplifier in common-emitter configuration.

Set the supply voltages to ±10v. Run an AC Sweep with the amplitude of Vsig set at 10mv. Print out a screen showing two plots, each with its own Y-axis:

• Voltage gain, V(out)/V(in) versus frequency on a decade frequency scale from 10Hz to 1MHz.

• Input impedance magnitude, |V(in)/I(Rs)|, over the same frequency range.

Amplifiers like these are often modeled in the "mid-band" region of frequency, where the amplifier gain is relatively flat, and at a maximum. Label on your plot the frequency range over which the "mid-band" region extends. Attach your plot in the space provided.

Attach Task 11.0.2 printout here:

11.0.3. Adjust the resistor values *in the amplifier part* of Fig. P11-3 (leave the source and the load resistances unchanged) to make the mid-band voltage gain as large as you can. When you think you have maximized the mid-band voltage gain, enter your new resistance values in the Data Table.

Data Table for Task 11.0.3 (Ω).

Rb =		Re1 =	
Rc =		Re2 =	

With your new resistance values from the Data Table, run an AC Sweep with the amplitude of Vsig set at 10mv. Print out a screen showing two plots, each with its own Y-axis:

• Voltage gain, V(out)/V(in) versus frequency on a decade frequency scale from 10Hz to 1MHz.
• Input impedance magnitude, |V(in)/I(Rs)|, over the same frequency range.

Label on your plot the frequency range over which the "mid-band" region extends. Attach your plot in the space provided.

P11-6

Attach Task 11.0.3 printout here:

11.0.4. Create a PSPICE Schematic of the universal BJT amplifier in the emitter-follower configuration as shown in Fig. P11-4. You should be able to do this quickly and easily by renaming the BUBBLES on the source, load, and ground circuits in your earlier schematic so that the appropriate connections are made.

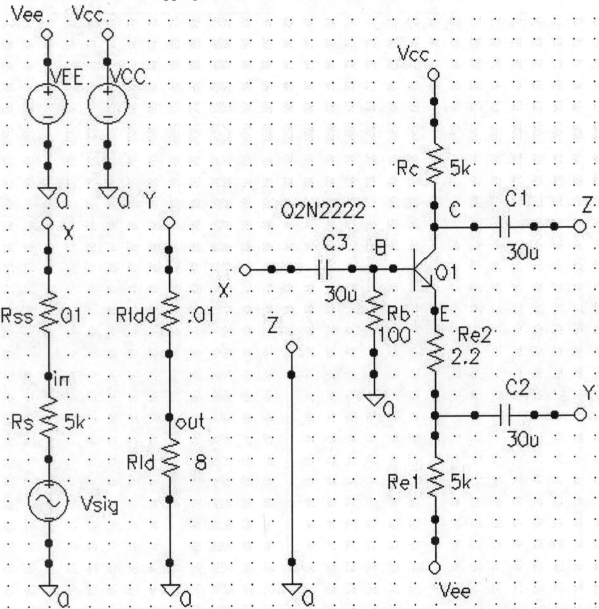

Fig. P11-4. Universal BJT amplifier circuit in the emitter-follower configuration.

Note that this amplifier is driving a low-impedance load from a high-impedance source.

Run an AC Sweep with the amplitude of Vsig set at 10mv. Print out a screen showing two plots, each with its own Y-axis:

P11-8

- Current gain, I(Rld)/I(Rs) versus frequency on a decade frequency scale from 10Hz to 1MHz.
- Input impedance magnitude, |V(in)/I(Rs)|, over the same frequency range.

Label on your plot the frequency range over which the "mid-band" region extends. Attach your plot in the space provided.

Attach Task 11.0.4 printout here:

11.0.5. Adjust the resistor values in the amplifier part of Fig. P11-4 (leave the source and the load resistances unchanged) to make the mid-band current gain as large as you can. When you've maximized the current gain as best you can, enter your resistor values in the Data Table.

Data Table for Task 11.0.5 (Ω).

Rb =		Re1 =	
Rc =		Re2 =	

Run an AC Sweep with the amplitude of Vsig set at 10mv. Print out a screen showing two plots:
• Current gain, I(Rld)/I(Rs), versus frequency on a decade frequency scale from 10Hz to 1MHz.
• Input impedance magnitude, |V(in)/I(Rs)|, over the same frequency range.

Label on your plot the frequency range over which the "mid-band" region extends. Attach your plot in the space provided.

Attach Task 11.0.5 printout here:

Task 11.1. Common-emitter amplifier measurements

11.1.1. Construct the common-emitter amplifier in Fig. P11-3, using standard resistors as close as possible to those in the figure. Supply Vcc and Vee at ±6v. Do not use any resistors for Rss and Rldd; just replace them with wires. Supply a .05v amplitude sine-wave for the input.

11.1.2. Consult your lab instructor and choose one of the methods below to measure the transfer function of your circuit. Circle your choice.

Transfer function measurement methods for Task 11.1.2.

	Method	Workbook section
1	*Virtual Bench* Oscilloscope	M4.1
2	*LabVIEW* Single-frequency transfer function	M4.2.1
3	*LabVIEW* Broad-band transfer function	M4.2.2

11.1.3. Draw your measurement circuit here. Show all analog input and output channel connections:

Measurement circuit:

11.1.4 While taking measurement data, you will need to make sure your frequency range covers the following features:

Transfer function part:	Feature(s)
Magnitude	Cover the "mid-band" region and about one decade either side, if possible.
Phase	

11.1.5. Measure the magnitude and phase of the transfer function over the required frequency range. If you are using method 1 or 2, fill in the appropriate columns in the Data Table. If you are using method 3, print out the front panel showing your results and attach it to this page, over the table and graph.

Data Table for Task 11.1.5

Freq. (Hz)	Virtual Bench Scope measurements		Virtual Bench calculations / LabVIEW measurements	
	Input Voltage Amplitude (v)	Output Voltage Amplitude (v)	|H|	Phase of H (deg)

Plot the magnitude and phase (two curves) from the Data Table on the graph below. Put the magnitude axis on the left and the phase axis on the right. Label all axes.

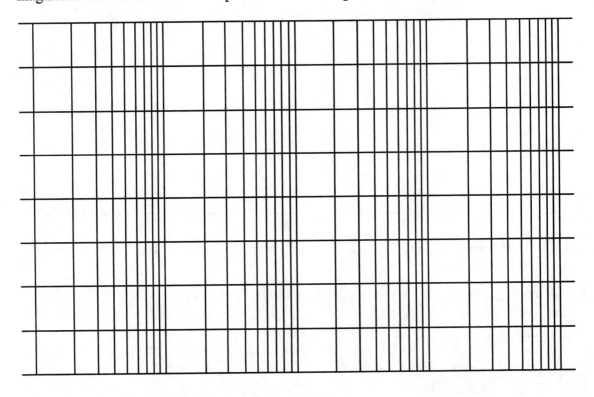

Task 11.2. Current Gain vs. frequency for the emitter-follower amplifier.

11.2.1. Review Section M5-4 in this workbook on measuring trans-impedance. Pay special attention to how a trans-impedance measurement on a circuit with a known load resistor at the output can be used to measure current gain.

11.2.2. Build the emitter-follower amplifier shown in Fig. P11-4. Ignore the very small (0.01 Ω) resistors. Use standard resistors with values as close as possible to those in Fig. P11-4.

11.2.3. Select a frequency well within the mid-band region. Use either the *Virtual Bench* Oscilloscope, or the modified Single-frequency Impedance VI as described in Section M5-4 to measure the mid-band current gain. Print out your instrument screen and attach it in the space provided.

11.2.4. Calculate the current gain from the screen data and the known values of the measuring and the load resistances. Show your calculation here, and enter the result:

Mid-band current gain = _____

Attach Task 11.2.3 printout here:

Project 12: The Source-follower FET Amplifier

Introduction

For every configuration of the universal BJT amplifier, there exists an equivalent single stage FET amplifier. Rather than explore the different FET amplifier configurations, this project will focus on one particularly simple and useful one, the FET source follower. Its primary uses are as an input stage in a multi-stage amplifier requiring a very high input impedance and as an output stage capable of supplying more current than might otherwise be available to a low-impedance load. Thus, it excels in applications where current gain and high input impedance are paramount and voltage gain unimportant, similar to the BJT emitter follower.

Source follower circuit

The basic source-follower circuit is described in Section 5.8 of Sedra and Smith, or Section 5.6 of Hambly. In the source-follower shown in Fig. P12-1, several capacitors and resistors have been eliminated from the universal FET amplifier circuit of Sedra and Smith.

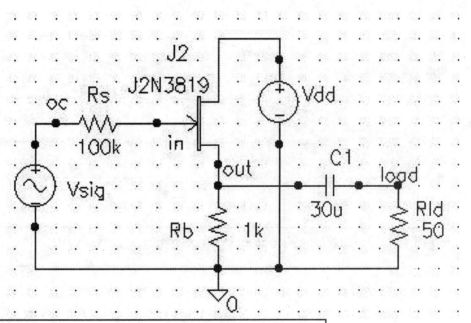

Fig. P12-1. Simplified source-follower circuit.

The signal source used for illustration in Fig. P12-1 is a particularly troublesome one, with a source impedance of 100 KΩ. If this source were to drive the 50Ω load shown, only a tiny fraction of its open circuit voltage (at node **oc**, in the diagram) would appear across the load. The source follower combines a very large input impedance with significant current gain. In this case, the voltage gain is substantially less than 1. However, in many applications the 50Ω load shown here is the input impedance of a BJT amplifier stage with high enough voltage gain to more than compensate for the low voltage gain of the source follower. These two amplifiers working together provide both high gain and very high input impedance, a combination which is hard to achieve with any single-stage amplifier.

Using the simple circuit in Fig. P12-1 as a high-impedance front-end for a multi-stage amplifier exacts some penalties. For example, it's rather unlikely that a circuit built exactly according to Fig. P12-1 specifications will perform exactly as simulated in PSPICE. The reason for this is that the gate voltage required to produce a particular value of drain to source current is not a well-controlled parameter in the manufacture of FET's. It will vary from device to device. Fig. P12.2 shows a simulation of the J2N3819's output characteristics.

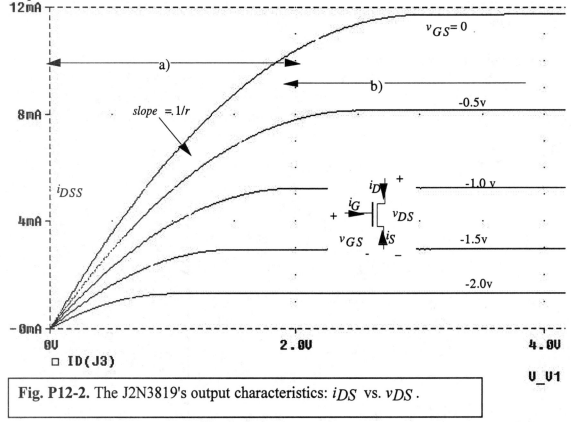

Fig. P12-2. The J2N3819's output characteristics: i_{DS} vs. v_{DS}.

The result of this variability in gate-to-source voltage v_{GS}, is that, qualitatively, the characteristics of your FET will look like Fig. P12-2, only the v_{GS} associated with each curve will have somewhat different values. This means v_{GS} has to be adjusted in each source-follower circuit to account for these variations. If the FET fails and is replaced, the resistor or supply voltage values may have to be changed to provide the right v_{GS} for the new FET.

Tasks
Task 12.0. Preparation
12.0.1. Review the above reference material on FET's and source-follower amplifiers.

12.0.2. In the source-follower amplifier, the output circuit of the FET acts as a current source, with i_{DS} essentially independent of v_{DS}. This means the FET is operating in the region labeled b) in Fig. P12-2, to the right of the dotted line.

Select the values of Rb and Vdd in Fig. P12.1 required to achieve a given DC bias point for the source-follower amplifier: Suppose your desired bias point is v_{GS} = -1.5v and v_{DS} = 3 v. Show your calculations for answering the following questions:

a) What is the DC bias current through Rb?

b) If the signal source Vsig has zero DC offset, what value for Rb do you need to get v_{GS} = -1.5v at the bias point?

c) Finally, what is the minimum value of Vdd required so that v_{DS} > 3 v in this circuit?

12.0.3. Using the values you calculated in Task 12.0.2 for Rb and Vdd, simulate with PSPICE the source-follower circuit in Fig. P12-1. Print out a single screen showing these plots on a decade frequency scale covering the range 100Hz < f < 1MHz
- Current gain at the load resistor Rld (log scale on Y-axis)
- Voltage gain at the load resistor Rld (log scale on Y-axis)
- Input impedance magnitude (log scale on Y-axis)

Use two Y-axes for the plot: one for the voltage and current gains, another for the input impedance. Note that this amplifier has a voltage gain substantially less that one but allows the very high impedance signal source to drive a low-impedance load with large current gain over a wide range of frequencies.

Attach the printout in the space provided.

Attach Task 12.0.3 printout here:

Task 12.1. Characterizing an individual FET and designing a source-follower around the results

12.1.1. Use the Curve Tracer VI from Chapter M4 to obtain a set of output characteristic curves for the FET you will use to build your source-follower amplifier. If *LabVIEW* is not installed on your system your laboratory may have a free-standing curve tracer you can use for this purpose. Set up this virtual instrument so that you get a set of curves similar to Fig. P12-2, with about half the horizontal axis occupied by the region labeled b). Print out a copy of the Curve tracer screen and attach it in the space provided. Draw a line on your printout *equivalent* to the line separating the regions a) and b) in Fig. P12-2 and label the regions.

12.1.2. Choose a dc operating point near the middle (horizontally and vertically) of region b). Label this point Q on your printout. Answer the following questions:

What is the required dc gate-to-source voltage to achieve this operating point? _____ v.

What is the dc bias current that must flow through Rb? _____ A.

What is the exact value you need to use for Rb to operate your FET at the desired dc operating point, Q? Rb = _____ Ω.

What is the nearest standard resistor value to the above value? _____ Ω

To get the required voltage drops across Rb and the FET for operation at the point Q, what voltage should you apply at the dc source, Vdd.?

Vdd = _____ v.

Attach Task 12.1.1 printout here:

Task 12.2. Testing the source-follower design

12.2.1. Construct the source-follower of Fig. P12-1, using the standard value for Rb you determined in Task 12.1.2. However you generate the input signal for the current gain measurement, connect a 100K resistor in series for Rs. [The analog outputs of your DAQ board have a much lower source impedance than 100KΩ, but for this measurement, you need to simulate a high impedance source.] This series resistance can double as the measuring resistor for the trans-impedance measurement.

Set Vdd to the value you determined in Task 12.1.2. Use your bench DC power supply.

12.2.2. Use the *Virtual Bench* Oscilloscope or the Single-frequency Impedance VI from Section M5.3, modified for trans-impedance and current gain measurements as in Section M5.4, to measure current gain vs. frequency. Cover at least the frequency range 100 Hz $< f <$ 100 KHz, using a logarithmic (decade) scale for frequency.

Enter your measuring and your load resistance values here:

Rm (=Rs)	
Rload	

Draw your measurement circuit here. Show all analog input and output channel connections. Show power supply voltage connections.

Complete the Data Table below for this measurement.

Data Table for Task 12.2.2

Freq. (Hz)	Virtual Bench Oscilloscope only				Either method		
	Math Channel Amplitude	Output voltage Channel Amplitude	Input Current Amplitude I_{in}	Output Current Amplitude I_{out}	Trans-Impedance magnitude	Current gain	Phase shift I_{out} vs. I_{in}

Plot current gain and phase shift on the graph below. Label all axes.

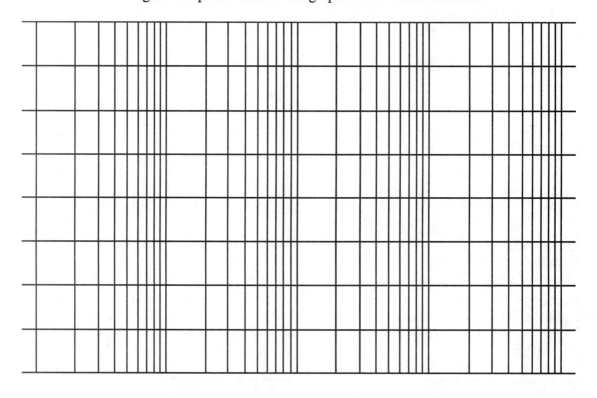

Project 13: Voltage-controlled Gain and Amplitude Modulation

Introduction

In this project, you will explore some fundamental concepts of amplitude modulation using a simple op-amp circuit. Modulation allows you to put information on a carrier wave that has a frequency spectrum that propagates well in some medium, such as free space, coaxial cable, or optical fiber. In typical amplitude modulation (AM) a low-frequency (f_m) signal controls the amplitude of a carrier wave of much higher frequency (f_c).

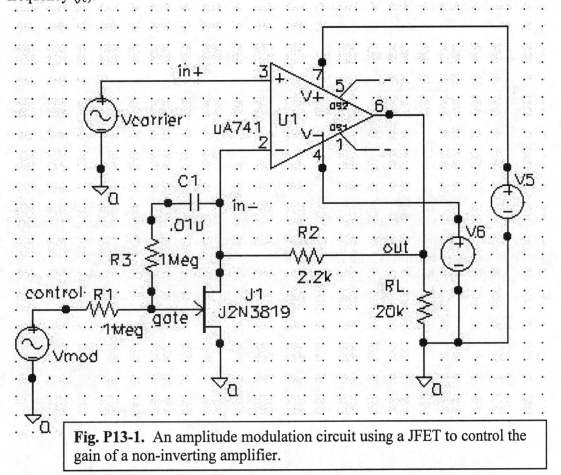

Fig. P13-1. An amplitude modulation circuit using a JFET to control the gain of a non-inverting amplifier.

The resistor R2 and the FET in Fig. P13-1 make up the voltage divider network for a non-inverting amplifier. If the voltage at node **in-** is small, the FET behaves as a resistor whose value is controlled by the gate voltage at node **gate**. Then, the input voltage to the amplifier at node **in+,** a sine wave of frequency f_c, is amplified by an amplifier whose gain depends on time with the form

$$G(t) = G_0 + g\sin(2\pi f_m t) . \qquad (\text{P13-1})$$

Tasks
Task 13.1. Simulation
13.1.1. Create a PSPICE Schematic like Fig. P13-1.

13.1.2. Set up the voltage sources in your schematic as for a Transient analysis shown in the Table:

Source:	Output form:
V5	Vcc = 10v DC
V6	Vee = -10v DC
Vcarrier	$v_c(t) = 1\sin(2\pi 10000t)$
Vmod	$v_m(t) = -1 + 0.4\sin(2\pi 2000t)$

See the Appendix for help in setting up sine wave sources in PSPICE.

13.1.3. Run a Transient analysis covering two cycles of $v_m(t)$, or about 1 ms. In PROBE, create three separate plots on a single screen. Show V(in+), V(control) and V(out) in these plots. Observe how V(out) is a sine wave at 10KHz, whose amplitude is modulated, periodically, at the lower frequency of 2 KHz. Print out a copy of this screen and attach it in the space provided.

13.1.4. Without changing any PROBE settings from Task 13.1.3, check Fourier under **Plot->X-axis->Processing Option**s. This will Fourier Transform the three waveforms. Adjust the X-Axis range so that only frequencies up to 15 KHz are shown. Observe how the spectrum of the AM output consists of a carrier at 10KHz, plus two satellites (called sidebands) at 12KHz and 8KHz, the sum and difference frequencies for f_c and f_m. Print out a copy of this screen and attach it in the space provided.

Attach Task 13.1.3 printout here

Attach Task 13.1.4 printout here:

Task 13.2. Determine JFET small signal behavior.

Review Chapter M7 on Small-signal parameter measurements. Select a J2N3819 JFET to use as the voltage-controlled resistor in your AM circuit. Consider it in the common-source configuration. Measure its Y-parameters at the following three Q-points:

Q-point settings:

General voltage:	Device voltage	Q-point 1	Q-point 2	Q-point 3
v_{in}	V_{GS}	-1 volt	-1.5 volt	-0.5 volt
v_{out}	V_{DS}	0.5 volt	0.5 volt	0.5 volt

The values for the small-signal admittance parameter y_{22} are the reciprocals of the small signal resistance between the drain and the source of the JFET. Complete the Data Table below by entering these resistances, and calculating the resulting gains for your non-inverting amplifier.

Data Table for Task 13.2

Q-pt	V_{GS}	y_{22}	$r = 1/y_{22}$	Amp. Gain
3	-0.5 v			
1	-1.0 v			
2	-1.5 v			

Among these three Q-points, what is the largest percent variation in amplifier gain?

Answer = _____ %. This percent figure is an estimate of the modulation index of the output waveform. Higher modulation indices are generally, but not always desirable.

Task 13.3. AM Circuit Demo

WARNING! This task requires the use of a second, free-standing signal generator to supply the carrier voltage. If you cannot obtain a second function generator, do not undertake this task.

13.3.1. Build the AM circuit, using standard components with values as close as possible to those in Fig. P13-1. Set up the DC power supply voltages, carrier and control voltage amplitudes and DC levels, exactly as they were in your simulation: i.e. the Table in Task 13.1.2.

13.3.2. Capture and display the amplitude spectrum of **exactly** four cycles of the output of your AM circuit. Choose one of the following methods:

	Method	Workbook Reference
1	*Virtual Bench* Digital Spectrum Analyzer	M3.2
2	*LabView* Spectrum Analyzer VI	M3.3

P13-6

Print out a front panel of your instrument showing the time waveform and the amplitude spectrum and attach it in the space provided. Choose a sampling rate fast enough for you to see spectral components up to 15 KHz. Use a logarithmic or dB scale on the vertical axis of your amplitude spectrum plot. Identify on your printout the carrier frequency and the sidebands.

Attach Task 13.3.2 printout here:

13.3.3. Estimate the modulation index from your measured time waveform using the formula,

$$MI = \frac{(highest\_amplitude) - (lowest\_amplitude)}{highest\_amplitude} \times 100$$

Estimated Modulation Index = _____ %

Appendix: PSPICE Essentials

Introduction

Most universities have access to some form of the Windows version of PSPICE. It has been available as freeware to institutions of higher education for many years. There are many comprehensive textbooks available that cover all aspects of its operation. This Appendix will briefly summarize only those parts of this versatile software package that are necessary to simulate and analyze the circuits used as examples in the Project Chapters. The processes of putting parts on a Schematic and wiring them together are obvious enough that they will not be presented here.

A1. Analyses

The Project chapters in this workbook utilize only four of the Analyses available in PSPICE. For more detailed explanations, consult one or more of the references in the Reference Table below:

Reference:	Transient Analysis	DC Sweep Analysis	AC Sweep Analysis	Parametric Analysis
R. W. Goody, *MicroSim PSPICE for Windows*, 2nd ed., Prentice-Hall	Chs. 3 & 7	Chs. 2 & 4	Chs. 4 & 5	Ch. 6

1. Transient Analysis: This analysis produces time-domain plots in PROBE of any legal variable in the PSPICE Schematic. It works by solving in the time-domain the differential equations governing the voltage across and the current through each part on the Schematic. To produce an output from a circuit in the time-domain, you must have at least one voltage or current source whose attributes are also specified in the time domain. Once plotted, the resulting time waveforms can be Fourier transformed to yield frequency-domain information.
2. DC Sweep Analysis: This analysis produces plots of DC voltages and currents in the circuit as functions of another DC voltage and/or current.
3. AC Sweep Analysis: This analysis produces frequency-domain plots of the amplitudes and phases of voltages and currents. The horizontal axis in these plots is frequency.
4. Parametric Analysis: This analysis can be done in conjunction with the others. It allows you to vary a parameter in the circuit. The Parametric Analysis essentially causes the other analysis paired with it to be repeated for different values of the parameter being varied.

You select the analysis/analyses to be performed in the Analysis->Setup menu shown in Fig. A1.

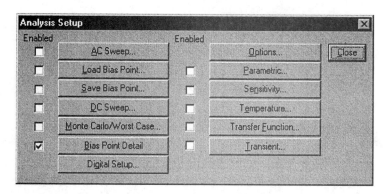

Fig. A-1.
The Analysis Setup Menu.

A1.1. Transient Analysis. Figure A-2 shows the Transient analysis setup menu. You need to set the **Final Time** at the time your analysis is to end. **Step Ceiling** can keep the program from running too long by limiting the total number of time steps allowed. The **Fourier Analysis** section at the bottom of the menu lets you calculate a Fourier Series for a periodic waveform. You must give the program the **Center Frequency** (the frequency of the fundamental harmonic) and the number of harmonics to be calculated, as well as the **Output Variables** for which you wan

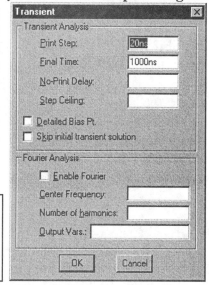

Fig. A-2.
Transient Analysis setup menu.

The resulting Fourier Series doesn't appear in PROBE, but at the end of the PSPICE output file. You get to this file by selecting **Analysis->Examine Output.**

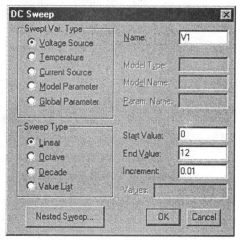

Fig. A-3.
DC Sweep setup menu

A1.2. DC Sweep Analysis. Figure A-3 shows the DC Sweep setup menu. In the figure, this menu is set up the way you will use it most often. It will cause the voltage source V1 to sweep from 0 to 12 volts in steps of 0.01 volt.

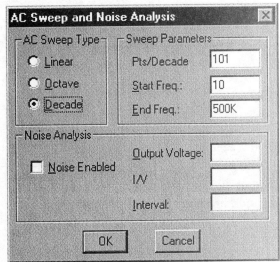

Fig. A-4.
AC Sweep setup menu.

A1.3. AC Sweep Analysis. Figure A-4 shows the AC Sweep setup menu. You must define the start and end frequencies for the frequency domain plot, and the type of scale. The Noise Analysis feature is beyond the scope of this workbook.

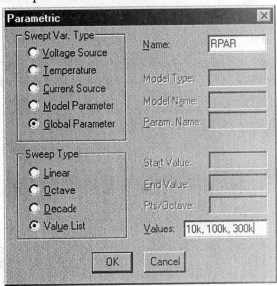

Fig. A-5.
Parametric Analysis setup menu.

A.1.4. Parametric Analysis. Figure A-5 shows the Parametric Analysis setup menu. In the figure, the value of a component will take on three values, 10k, 100k and 300k. The value of the particular component in the schematic diagram has been replaced by **{RPAR}**, and the name of the parameter, RPAR, has been added as a **NAME** in a part on the schematic called a **Parameter List**. Figure A-6 shows how the schematic looks in a parametric analysis where the value of a resistor is to be the parameter varied. Component values such as resistor or capacitor values are "**Global Parameters**" in Fig. A-5.

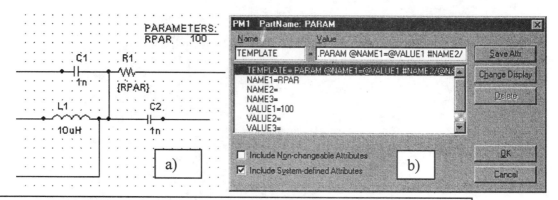

Fig. A-6. a) Schematic with the value of resistor R1 made a parameter, b) the Parameter List.

You "Get" the **Parameter List** like any other part, then double-click on it in Fig, A-6a) to open the menu in Fig. A-6b). The value you enter in Value1 in Fig. A-6b) will only be used if you disable **Parametric Analysis** in the main Analysis Setup menu, Fig. A-1. If you check the **Parametric...** box there, Fig. A-6 determines what values R1 will take on in the analysis.

A2. Some PSPICE signal sources

One of the parts libraries in PSPICE, called the **source.slb**, contains a wide variety of independent and dependent voltage and current sources. The projects in this workbook only utilize four of them: the independent voltage sources, VSRC, VPULSE, VPWL, and VSIN. Once you understand how these work, it is easy to learn to use the other sources.

A2.1. The simple voltage source, VSRC. This voltage source provides the following voltages:
- a constant DC level in a Transient analysis, set by its attribute, **TRAN**.
- a sine wave in an AC Sweep analysis, whose amplitude is set by the attribute, **AC**.
- a DC voltage in a DC analysis, set by the attribute **DC**.

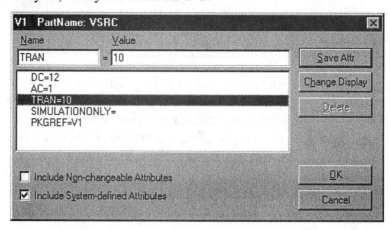

Fig. A-7. The Attributes of a simple voltage source, VSRC.

Figure A-7 shows the menu for setting the attributes for the VSRC source. In a Transient Analysis, this source would be a DC level of 10 volts, while in an AC sweep, it would be a sine wave of amplitude one volt, whose frequency would vary over the range specified in the AC Sweep setup menu.

A2.2. The pulsed voltage source, VPULSE

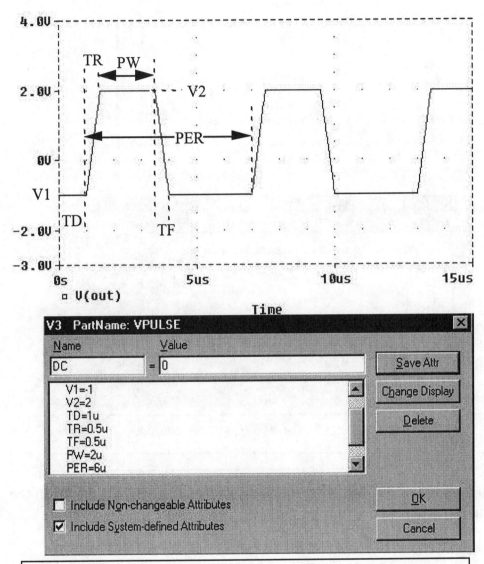

Fig. A-8. Time waveform and attributes of the VPULSE voltage source in PSPICE.

This voltage source provides various pulsed waveforms where up and down transitions are made between two voltage levels. Figure A-8 shows how its attributes relate to its time waveform. You can use this source to make pulse trains, square waves and triangle waves.

A2.3. The piecewise linear voltage source, VPWL

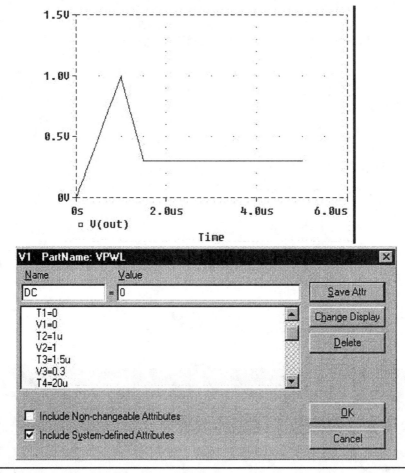

Fig. A-9. Waveform and attributes of the VPWL voltage source.

In principle, you could make any waveform you wanted with this voltage source. Its time waveform is a series of straight lines connecting the points (Vi,Ti) in order of the index, i. Figure A-9 shows its attributes and time waveform.

A2.4. The sine-wave source, VSIN

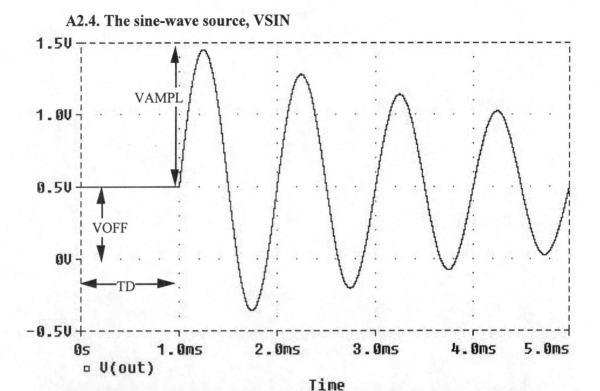

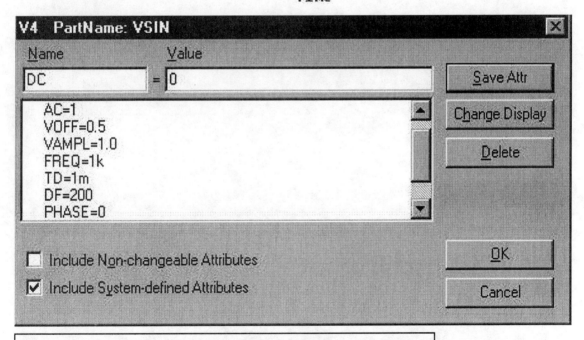

Fig. A-10. Time waveform and attributes of the VSIN source.

Use this source to make sine waves. You can set every parameter you see in the attribute list in Fig. A-10. The attribute, **DF**, is a damping factor and has the units of reciprocal seconds.